Everyday

Mathematics®

The University of Chicago School Mathematics Project

Skills Link

Cumulative Practice Sets
Student Book

Mc Graw Hill **Wright Group**

The McGraw·Hill Companies

Photo Credits

Cover—©Tim Flach/Getty Images, cover; Getty Images, cover, *bottom left*
Photo Collage—Herman Adler Design

www.WrightGroup.com

 **Wright Group**

Send all inquiries to:
Wright Group/McGraw-Hill
P.O. Box 812960
Chicago, IL 60681

ISBN 978-0-07-618786-7
MHID 0-07-618786-1

1 2 3 4 5 6 7 8 9 CPS 13 12 11 10 09 08 07

The *McGraw·Hill* Companies

Contents

Name _____ Date _____ Time _____

Practice Set 1

Use with or after
Lesson 1·2

SRB
6–9

Write your answers below or on another piece of paper.

Find the missing number.

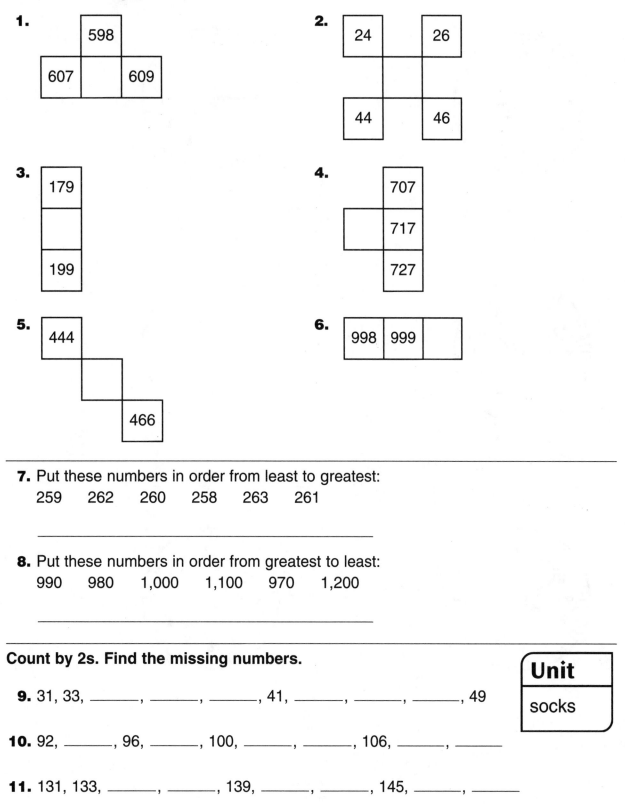

1.

	598	
607		609

2.

24		26
44		46

3.

179
199

4.

	707
	717
	727

5.

444

| | 466 |

6.

998	999	

7. Put these numbers in order from least to greatest:

259   262   260   258   263   261

_____

8. Put these numbers in order from greatest to least:

990   980   1,000   1,100   970   1,200

_____

Count by 2s. Find the missing numbers.

Unit

socks

9. 31, 33, _____, _____, _____, 41, _____, _____, _____, 49

10. 92, _____, 96, _____, 100, _____, _____, 106, _____, _____

11. 131, 133, _____, _____, 139, _____, _____, 145, _____, _____

1

Practice Set 2

**Use with or after
Lesson 1·4**

SRB
174

Write your answers below or on another piece of paper.

Record the time shown on each clock.

1.

_____

2.

_____

3.

_____

4.

_____

5.

_____

6.

_____

Practice Set 2 *continued*

Write your answers below or on another piece of paper.

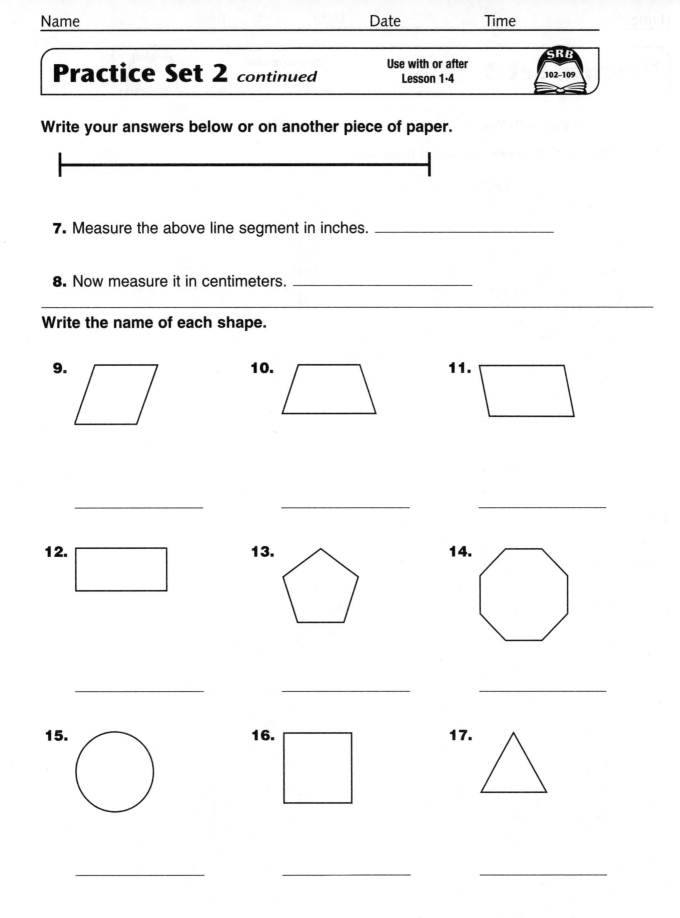

7. Measure the above line segment in inches. _____

8. Now measure it in centimeters. _____

Write the name of each shape.

9. _____

10. _____

11. _____

12. _____

13. _____

14. _____

15. _____

16. _____

17. _____

Practice Set 3

Use with or after Lesson 1·5

Write your answers below or on another piece of paper.

Use the tally chart to answer each question.

Ice Cream Favorites	
Kind of Ice Cream	**Number of Students**
Vanilla	卌
Strawberry	卌 //
Double Chocolate	卌 卌 //
Chocolate Chip Mint	////
Maple Nut	///

1. How many students like Double Chocolate ice cream best? _____

2. What is the favorite flavor? _____

3. What is the least-favorite flavor? _____

4. How many students altogether chose Double Chocolate or Chocolate

Chip Mint? _____

Write the number shown by the base-10 blocks.

5.

6.

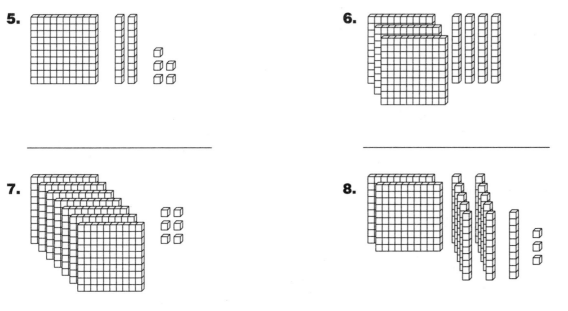

7.

8.

Practice Set 3 *continued*

Use with or after
Lesson 1·5

SRB
86-87

Write your answers below or on another piece of paper.

Use the bar graph to answer each question.

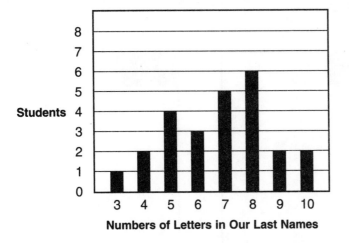

9. How many students have eight letters in their last names? _____

10. How many letters are in the *shortest* last name? _____

11. How many names are shown by the bar graph? _____

12. How many students have fewer than six letters in their last names? _____

13. How many letters does the *longest* name have? (This is called the *maximum.*)

_____

14. How many letters does the *shortest* name have? (This is called the *minimum.*)

_____

15. What is the *range* of the numbers of letters? _____

16. What is the *mode* of this set of data? _____

(*Hint:* If you don't remember what range and mode are, look them up in your *Student Reference Book.*)

Practice Set 4

Use with or after
Lesson 1·6

SRB
14 15

Write your answers below or on another piece of paper.

Make up your own name-collection box for each of the five numbers listed below. Include 10 different names for each number.

Example

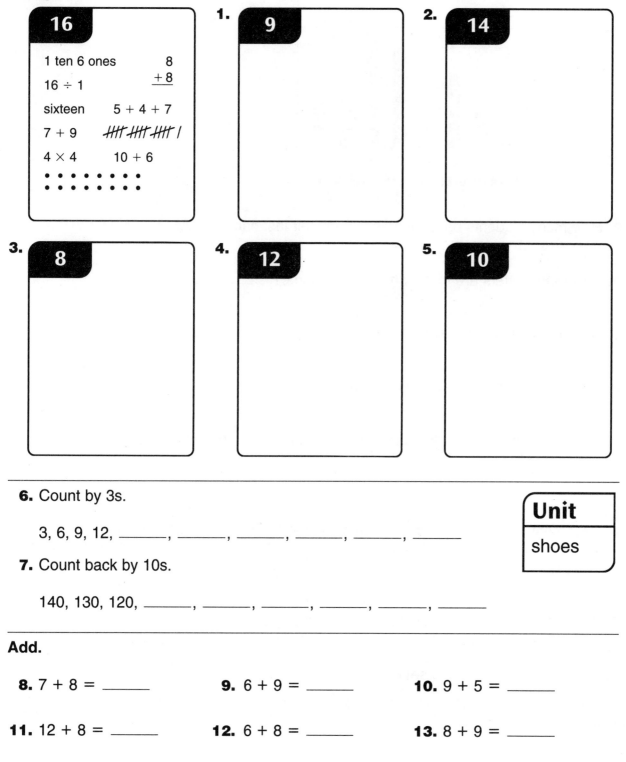

16

1 ten 6 ones	8
16 ÷ 1	+ 8
sixteen	5 + 4 + 7
7 + 9	╫╫ ╫╫ ╫╫ /
4 × 4	10 + 6

1. **9**

2. **14**

3. **8**

4. **12**

5. **10**

6. Count by 3s.

   3, 6, 9, 12, _____, _____, _____, _____, _____, _____

7. Count back by 10s.

   140, 130, 120, _____, _____, _____, _____, _____, _____

| Unit |
| shoes |

Add.

8. 7 + 8 = _____

9. 6 + 9 = _____

10. 9 + 5 = _____

11. 12 + 8 = _____

12. 6 + 8 = _____

13. 8 + 9 = _____

Practice Set 5

Write your answers below or on another piece of paper.

									0
1	2	3	4	5	6	7	8	9	10
11	12	13	14	15	16	17	18	19	20
21	22	23	24	25	26	27	28	29	30
31	32	33	34	35	36	37	38	39	40
41	42	43	44	45	46	47	48	49	50
51	52	53	54	55	56	57	58	59	60
61	62	63	64	65	66	67	68	69	70
71	72	73	74	75	76	77	78	79	80
81	82	83	84	85	86	87	88	89	90
91	92	93	94	95	96	97	98	99	100
101	102	103	104	105	106	107	108	109	110

1. Find 20 more than 84. _____

2. Find 16 more than 68. _____

3. Find 12 less than 32. _____

4. Find 35 less than 44. _____

5. Start at 0 and count by 3s along the *second* line of the number grid. Write the numbers from your count.

_____

6. Start at 41 and count by 3s along the *sixth* line of the number grid. Write the numbers from your count.

_____

7. Start at 81 and count by 6s along *two lines* of the number grid. Write the numbers from your count.

_____

Practice Set 6

Write your answers below or on another piece of paper.

Use your calculator to count by 10s. Find the missing numbers.

Example   40, _____, _____, 70, 80, _____, _____

   Press: ④ ⓪ ⊕ ① ⓪ ⊜ ⊜ ⊜ ⊜ ⊜ ⊜

   Display: 50, 60, 70, 80, 90, 100

1. 25, _____, 45, _____, _____, _____, 85, _____, _____, _____

2. 123, _____, 143, _____, _____, _____, _____, _____, _____, 213

Use your calculator to solve each problem.

3. The first English colony was established in the New World in 1607. The colonies united to demand freedom from England in 1776. How many years went by before the colonies demanded freedom?

_____

4. Marta read a book that was 45 pages long. Next, she read a book that was 82 pages long. Then she read a book that was 106 pages long. How many pages did Marta read in all?

_____

Write the missing numbers.

Unit
books

5. 10 = _____ + 4

6. 5 + _____ = 10

7. _____ + 7 = 10

8. 26 + _____ = 30

9. 50 = _____ + 43

10. 81 + _____ = 90

8

Practice Set 7

Write your answers below or on another piece of paper.

Estimate to answer *yes* or *no*.

1. You have $5.00. Do you have enough to buy a notebook for $3.99 and a pen for $1.55?

_____

2. You have $4.50. Do you have enough to buy two boxes of pencils that cost $2.10 each?

_____

3. You have $10.00. Do you have enough to buy crayons for $1.89, a backpack for $6.98, and paper clips for 79¢?

_____

4. You have $3.20. Do you have enough to buy a marker for $1.79 and a pad of paper for $1.49?

_____

Solve each problem.

5. Juana paid for a video that cost $6.59 with a $10.00 bill. How much change did she receive?

_____

6. Larry's lunch cost $3.25. Larry paid for his lunch with a $5.00 bill. How much change did he receive?

_____

Find an equal amount of money in the second list. Then write the letter that identifies that amount.

7. $\frac{1}{2}$ dime _____

A. dime

8. quarter _____

B. $0.50

9. $\frac{1}{10}$ dollar _____

C. $\frac{3}{4}$ dollar

10. $0.01 _____

D. penny

11. $\frac{1}{2}$ dollar _____

E. nickel

12. $0.75 _____

F. $\frac{1}{4}$ dollar

Practice Set 7 *continued*

Write your answers below or on another piece of paper.

Write =, <, or >.

13. $1.59 _____ $0.95

| = means *is equal to* |
| < means *is less than* |
| > means *is greater than* |

14. $7.52 _____ $4.75

15. $0.88 _____ $1.08

16. $6.65 _____ $5.66

17. $10.01 _____ $9.10   **18.** $0.75 _____ 75 cents

19. $1.11 _____ 111 pennies   **20.** 63 cents _____ $1.63

Use your calculator. Enter each amount of money. Then write the equivalent numeric value that is displayed on your calculator.

Example   Enter: 93¢   Display shows: 0.93

21. $0.08 _____   **22.** $1.59 _____

23. 98¢ _____   **24.** $6.57 _____

25. 3¢ _____   **26.** 59¢ _____

27. $2.43 _____   **28.** $0.79 _____

Draw coins to show each amount of money in two different ways.

Example   87¢

| Q Q Q | D D D D D |
| D P P | N N N N N P P |

29. 42¢   **30.** $0.35   **31.** 27¢

32. $0.54   **33.** 68¢   **34.** 76¢

Practice Set 8

Write your answers below or on another piece of paper.

Complete each Frames-and-Arrows diagram.

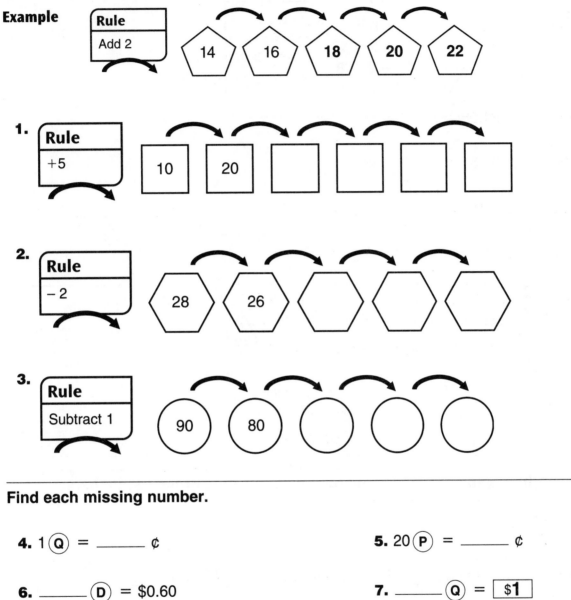

Example

Rule
Add 2

14 16 18 20 22

1.

Rule
+5

10 20 ☐ ☐ ☐ ☐

2.

Rule
− 2

28 26

3.

Rule
Subtract 1

90 80

Find each missing number.

4. 1 (Q) = _____ ¢

5. 20 (P) = _____ ¢

6. _____ (D) = $0.60

7. _____ (Q) = $1

8. 1 half-dollar = _____ ¢

9. _____ (Q) = 10 (N)

10. 1 quarter = _____ dimes and _____ nickels

11. _____ dimes and _____ pennies = 1 dollar

Practice Set 9

Write your answers below or on another piece of paper.

Write the number family for each Fact Triangle.

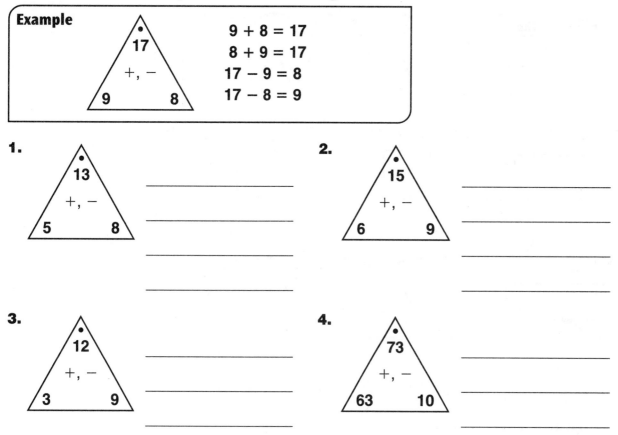

Example

△ 17 +, − 9 8

$9 + 8 = 17$
$8 + 9= 17$
$17 − 9 = 8$
$17 − 8 = 9$

1. △ 13 +, − 5 8 _____ _____ _____

2. △ 15 +, − 6 9 _____ _____ _____

3. △ 12 +, − 3 9 _____ _____ _____

4. △ 73 +, − 63 10 _____ _____ _____

Write the addition and subtraction number family for each group of numbers.

Example 7, 14, 7 $7 + 7 = 14$
                        $14 − 7 = 7$

5. 8, 8, 16 _____

6. 9, 9, 18 _____

7. 20, 20, 40 _____

8. 28, 14, 14 _____

9. 30, 30, 60 _____

10. 36, 18, 18 _____

Practice Set 9 *continued*

Use with or after
Lesson 2·1

SRB
200 201

Write your answers below or on another piece of paper.

Use two rules for each set of Frames and Arrows. Write the numbers for the empty frames.

Example

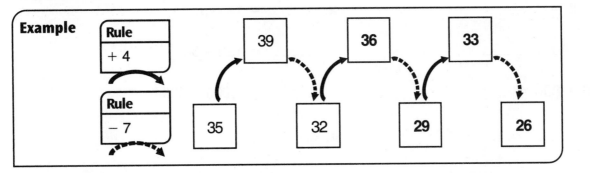

11.

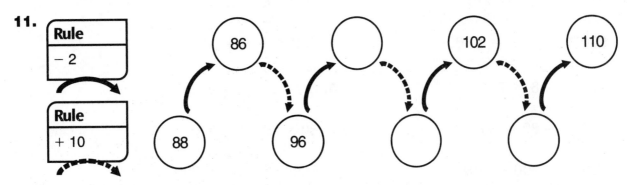

12.

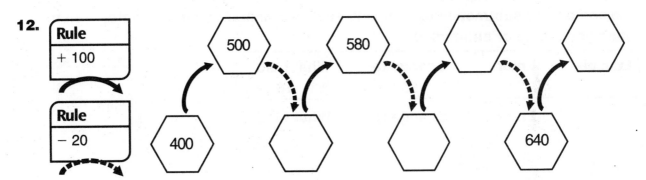

13.

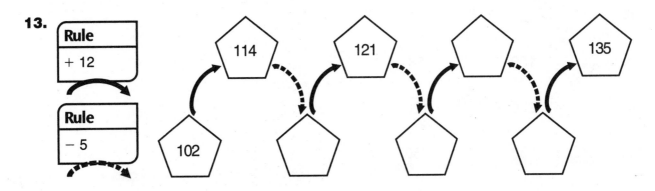

13

Practice Set 10

Use with or after
Lesson 2·2

SRB
254–255

Write your answers below or on another piece of paper.

Find each missing number.

1. 6 + 8 = _____

   60 + 80 = _____

   600 + 800 = _____

2. _____ = 12 – 5

   _____ = 120 – 50

   _____ = 1,200 – 500

3. 9 + _____ = 13

   90 + _____ = 130

   900 + _____ = 1,300

4. 4 = 9 – _____

   40 = 90 – _____

   400 = 900 – _____

5. _____ – 8 = 7

   _____ – 80 = 70

   _____ – 800 = 700

6. 8 = _____ – 9

   80 = _____ – 90

   800 = _____ – 900

Use addition or subtraction to complete each problem on your calculator. Tell how much you added or subtracted.

Example	Enter 34	Change to 50	What I did + 16
	Enter	**Change to**	**What I did**
7.	90	72	_____
8.	22	50	_____
9.	100	58	_____
10.	200	120	_____
11.	130	250	_____
12.	900	400	_____

14

Practice Set 10 *continued*

Use with or after
Lesson 2·2

SRB
14 15

Write your answers below or on another piece of paper.

Circle the names that DO NOT belong in each name-collection box.

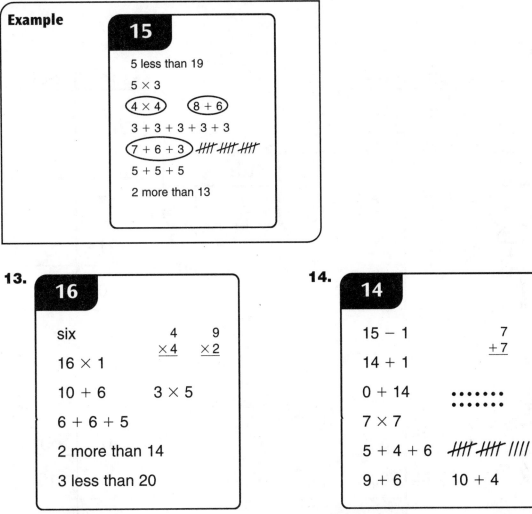

Example

15

5 less than 19

5 × 3

(4 × 4)   (8 + 6)

3 + 3 + 3 + 3 + 3

(7 + 6 + 3)   ╫╫ ╫╫ ╫╫

5 + 5 + 5

2 more than 13

13.

16

six   4   9
                  ×4   ×2

16 × 1

10 + 6   3 × 5

6 + 6 + 5

2 more than 14

3 less than 20

14.

14

15 − 1   7
                       +7

14 + 1

0 + 14   • • • • • • •
              • • • • • • •

7 × 7

5 + 4 + 6   ╫╫ ╫╫ ////

9 + 6   10 + 4

15.

12

12 + 1   3
                    ×4

20 − 8

3 + 10   1 × 12

5 + 8   2 × 6

• • •   ╫╫ ///
• • •
• • •

16.

20

twenty   4 × 5

╫╫ ╫╫ ╫╫ ////

13 + 8

20 + 0

5 + 6 + 9   10 + 10

20 × 0

8 + 7 + 7

Practice Set 11

Use with or after
Lesson 2·3

SRB
202–204

Write your answers below or on another piece of paper.

Find each missing number.

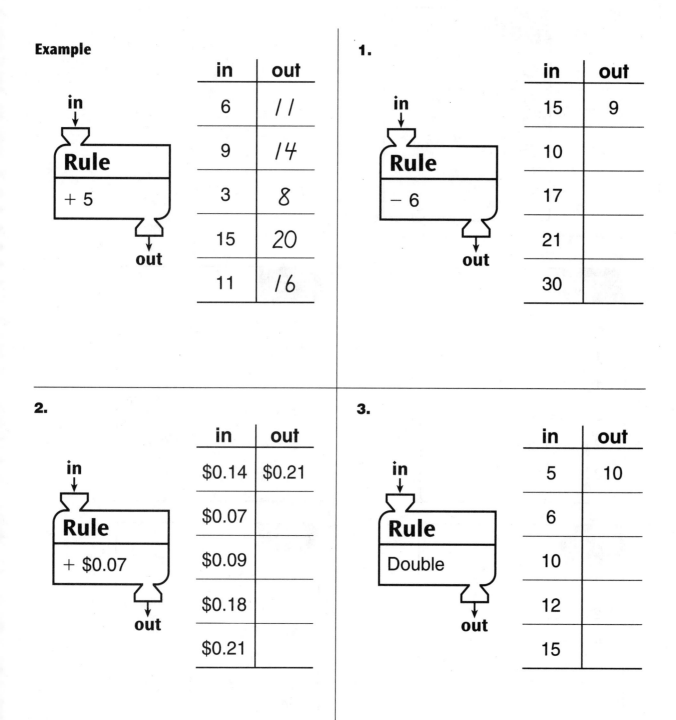

Example

in
Rule
+ 5
out

in	out
6	11
9	14
3	8
15	20
11	16

1.

in
Rule
− 6
out

in	out
15	9
10	
17	
21	
30	

2.

in
Rule
+ $0.07
out

in	out
$0.14	$0.21
$0.07	
$0.09	
$0.18	
$0.21	

3.

in
Rule
Double
out

in	out
5	10
6	
10	
12	
15	

Practice Set 12

Use with or after
Lesson 2·6

SRB
254 258

Write your answers below or on another piece of paper.

Choose one of the diagrams to help you solve each number story.

1. Bedelia picked 23 flowers Tuesday. She picked more flowers Wednesday. She picked a total of 47 flowers. How many flowers did she pick Wednesday?
   Answer the question:

   _____

   Number model: _____

Quantity

Quantity	
	_____
	Difference

2. Sara spelled 83 words correctly at this year's spelling contest. Last year she spelled only 47 words correctly. What is the difference between her two scores?
   Answer the question:

   _____

   Number model: _____

Total	
Part	**Part**

3. Larissa made 67 paper birds for a craft fair. Then she made 38 paper insects. How many objects did she make in all?
   Answer the question:

   _____

   Number model: _____

   Change

Start		End
	_____	

Practice Set 12 *continued*

Use with or after
Lesson 2·6

SRB
18–21

Write your answers below or on another piece of paper.

Count by 10s. Find the missing numbers.

4. 400; 410; _____; 430; _____; _____; 460; _____; _____; _____; 500

5. 1,010; 1,020; _____; 1,040; _____; _____; 1,070; _____; _____; 1,100

6. 3,225; 3,235; _____; _____; 3,265; _____; _____; _____; 3,305; _____

7. 8,712; 8,722; _____; 8,742; _____; _____; 8,772; _____; _____; 8,802

8. 3,218; 3,228; _____; _____; 3,258; _____; _____; _____; 3,298; _____

Tell what the underlined digit stands for in each number.

Example 9,$\underline{6}$13   6 hundreds, or 600

9. $\underline{2}$,917 _____

10. 3,04$\underline{6}$ _____

11. 8$\underline{5}$1 _____

12. 8,$\underline{0}$46 _____

13. $\underline{5}$,425 _____

14. $\underline{1}$4,523 _____

15. 6,79$\underline{1}$ _____

16. 4,3$\underline{8}$0 _____

17. 6$\underline{3}$,941 _____

Write the addition and subtraction number family for each group of numbers.

Example 5, 17, 22   5 + 17 = 22; 17 + 5 = 22
22 − 5 = 17; 22 − 17 = 5

18. 28, 9, 37 _____

19. 50, 30, 80 _____

20. 6, 57, 63 _____

21. 60, 8, 52 _____

22. 70, 90, 160 _____

23. 400, 500, 900 _____

18

Practice Set 13

Write your answers below or on another piece of paper.

Estimate first. Then use the partial-sums addition method to add.

Example

100s	10s	1s
4	6	7
+	1	8
4	0	0
	7	0
	1	5
4	8	5

1. 345
     + 69

Ballpark estimate:

470 + 20 = 490

Ballpark estimate:

_____

2.   38
      + 45

3.   75
      + 129

Ballpark estimate:

_____

Ballpark estimate:

_____

Practice Set 14

Write your answers below or on another piece of paper.

Estimate first. Then use the trade-first subtraction method to subtract.

Example

100s	10s	1s
1	12	
2̸	2̸	5
− 1	7	3
	5	2

1.   305
    − 69

Ballpark estimate:

225 − 175 = 50

Ballpark estimate:

_____

2.   138
    − 45

3.   275
    − 129

Ballpark estimate:

_____

Ballpark estimate:

_____

Practice Set 14 *continued*

Use with or after
Lesson 2·8

SRB
50 51

Write your answers below or on another piece of paper.

Find each missing number.

4. $1 = _____ (D)

5. $0.42 = _____ (P)

6. _____ half-dollars = $2.00

7. _____ (N) = 4 (D)

8. 75¢ = _____ (Q)

9. _____ (D) = 1 half-dollar

10. _____ (N) = 35¢

11. $0.70 = _____ (D)

Add or subtract.

Unit
fish

12. 52 + 79 = _____

13. 98 − 46 = _____

14. 14 + 24 + 36 = _____

15. 81 − 49 = _____

16. 26 + 91 = _____

17. 65 − 28 = _____

18. 104
    − 67

19.   26
    + 13

20.   79
    − 24

21.   57
    + 26

Count by 100s. Find the missing numbers.

22. 1,000; 1,100; _____; 1,300; _____; _____; 1,600; _____; _____; 1,900

23. 2,450; 2,550; _____; _____; 2,850; _____; 3,050; _____; _____; 3,350

24. 7,304; 7,404; _____; _____; _____; 7,804; _____; 8,004; _____; _____

25. 5,416; _____; _____; 5,716; _____; _____; _____; 6,116; 6,216; 6,316

26. 2,883; _____; _____; 3,183; 3,283; _____; _____; _____; 3,683; _____

Practice Set 15

Use with or after Lesson 2·9

SRB
256–257

Write your answers below or on another piece of paper.

Use one of the diagrams below to help you solve each problem.

Total		
Part	Part	Part

Total			
Part	Part	Part	Part

1. Samuel bought presents for 40 cents, 50 cents, 60 cents, and 70 cents. How much money did he spend in all?

_____

Check: Does my answer make sense? _____

2. Trini rode her bicycle 12 miles Friday. She rode 14 miles Saturday and 15 miles Sunday. How many miles did she ride in all?

_____

Check: Does my answer make sense? _____

3. Jon, Dave, and Kevin collected rocks at the beach. Each boy collected 25 rocks. How many rocks did the boys collect in all?

_____

Check: Does my answer make sense? _____

4. The Torrey family was on vacation. One day, they spent $140 for a motel room, $130 for meals, and $200 at a park. How much money did they spend that day?

_____

Check: Does my answer make sense? _____

Practice Set 15 *continued*

Use with or after
Lesson 2·9

SRB
50 51

Write your answers below or on another piece of paper.

Find the missing numbers for each Addition and Subtraction Puzzle.

Example

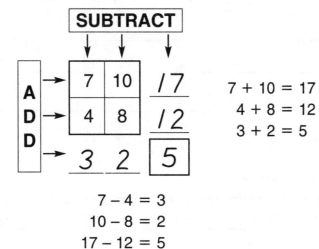

	SUBTRACT		
	7	10	*17*
ADD	4	8	*12*
	3	*2*	5

7 + 10 = 17
4 + 8 = 12
3 + 2 = 5

7 − 4 = 3
10 − 8 = 2
17 − 12 = 5

5.

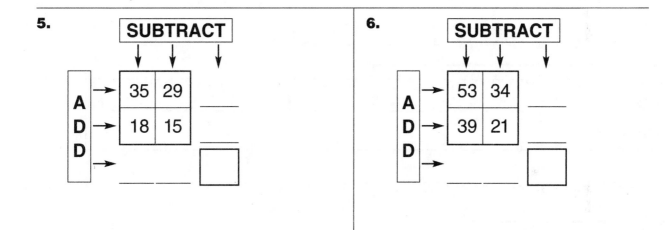

	SUBTRACT		
ADD	35	29	
	18	15	

6.

	SUBTRACT		
ADD	53	34	
	39	21	

7.

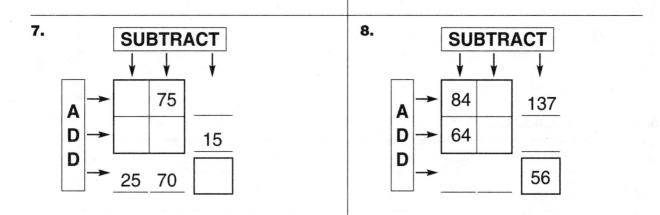

	SUBTRACT		
ADD		75	
			15
	25	70	

8.

	SUBTRACT		
ADD	84		137
	64		
			56

23

Practice Set 16

**Use with or after
Lesson 3-2**

SRB
143–145

Write your answers below or on another piece of paper.

Measure each line segment to the nearest $\frac{1}{4}$ inch.

1.

```
0    1    2    3    4    5    6
inches
```

_____

2.

```
0    1    2    3    4    5    6
inches
```

_____

3.

```
0    1    2    3    4    5    6
inches
```

_____

4.

```
0    1    2    3    4    5    6
inches
```

_____

5.

```
0    1    2    3    4    5    6
inches
```

_____

Practice Set 16 *continued*

Use with or after
Lesson 3·2

SRB
54 55

Write your answers below or on another piece of paper.

Find the missing number for each Fact Triangle. Then write the family of facts for that triangle.

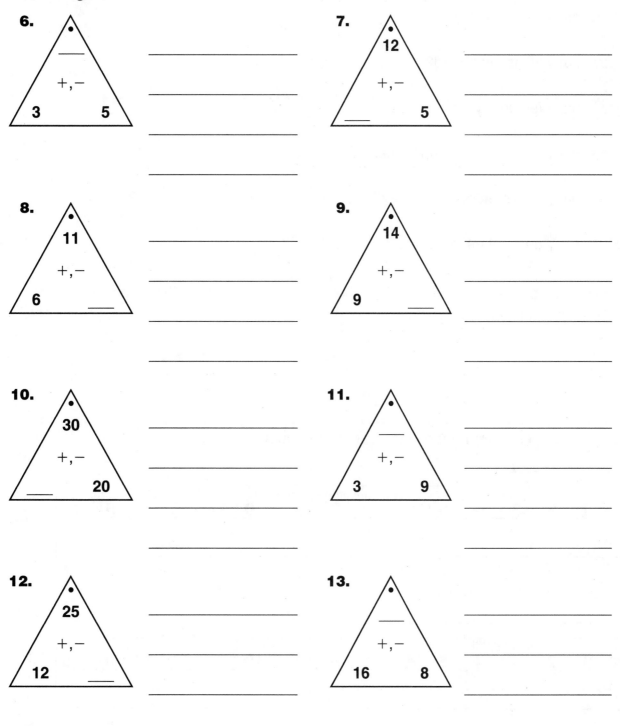

6. _____ +,− 3 5

7. 12 +,− _____ 5

8. 11 +,− 6 _____

9. 14 +,− 9 _____

10. 30 +,− _____ 20

11. _____ +,− 3 9

12. 25 +,− 12 _____

13. _____ +,− 16 8

Practice Set 17

Write your answers below or on another piece of paper.

Use measuring tools if you need help answering these questions:

1. How many inches in 1 foot? _____

2. How many inches in 1 yard? _____

3. How many feet in 1 yard? _____

4. How many inches in your tape measure? _____

5. How many centimeters in 1 meter? _____

6. How many decimeters in 1 meter? _____

7. Draw a line segment that you think is about 7 centimeters long. Then measure it to see how long it actually is.

Measure each line segment to the nearest centimeter.

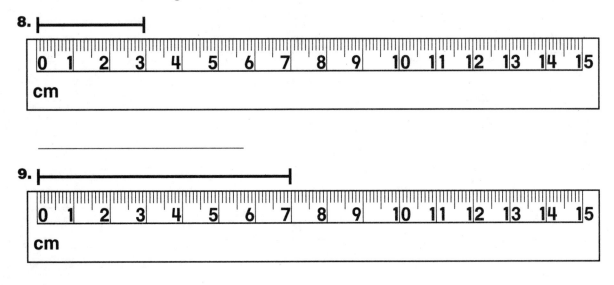

Practice Set 18

Use with or after
Lesson 3·5

SRB
150 151

Write your answers below or on another piece of paper.

Find the perimeter of each figure.

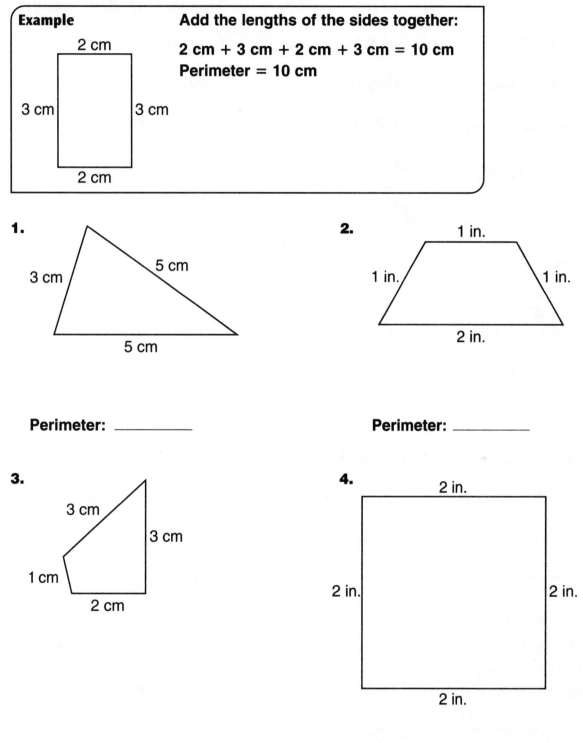

Example

2 cm

3 cm   3 cm

2 cm

Add the lengths of the sides together:

2 cm + 3 cm + 2 cm + 3 cm = 10 cm
Perimeter = 10 cm

1.

3 cm

5 cm

5 cm

Perimeter: _____

2.

1 in.

1 in.   1 in.

2 in.

Perimeter: _____

3.

3 cm

3 cm

1 cm

2 cm

Perimeter: _____

4.

2 in.

2 in.   2 in.

2 in.

Perimeter: _____

Practice Set 18 *continued*

**Use with or after
Lesson 3·5**

SRB
50–51
250–253

Write your answers below or on another piece of paper.

Find each sum or difference.

5. 7 + 4 = _____

17 + 4 = _____

27 + 4 = _____

37 + 4 = _____

47 + 4 = _____

6. 14 − 6 = _____

24 − 6 = _____

34 − 6 = _____

44 − 6 = _____

54 − 6 = _____

7. 6 + 3 = _____

60 + 30 = _____

600 + 300 = _____

8. 8 − 2 = _____

80 − 20 = _____

800 − 200 = _____

9. 9 + 6 = _____

90 + 60 = _____

900 + 600 = _____

10. 18 − 9 = _____

180 − 90 = _____

1,800 − 900 = _____

Count by 100s. Find the missing numbers.

11. 1,200; 1,100; _____; 900; 800; _____; _____; 500; _____; 300

12. 5,630; 5,530; _____; _____; 5,230; 5,130; _____; 4,930; _____; _____

13. 2,807; _____; 2,607; _____; _____; _____; 2,207; _____; _____; 1,907

14. 7,659; _____; _____; 7,359; _____; _____; _____; 6,959; _____; _____

15. 5,312; 5,212; _____; _____; _____; 4,812; 4,712; _____; _____; _____

Solve each problem.

16. The distance from Dallas to Houston is 245 miles. The distance from Dallas to El Paso is 617 miles. How much farther is it from Dallas to El Paso than from Dallas to Houston?

17. On their vacation, the Baker family drove 376 miles from Phoenix to Los Angeles. Then the Bakers drove 387 miles to San Francisco. How many miles did they drive in all?

_____

_____

Practice Set 19

Write your answers below or on another piece of paper.

Write a number model for each rectangle. Then find the area.

Example

Number model: $3 \times 4 = 12$

Area = 12 square units

1. 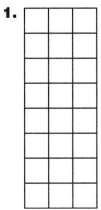   Number Model:

_____

Area: _____

2.   Number Model:

_____

Area: _____

3. 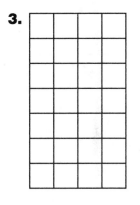   Number Model:

_____

Area: _____

4.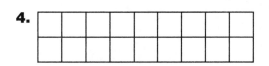

Number Model:

_____

Area: _____

Practice Set 19 *continued*

Use with or after
Lesson 3·7

SRB
150 151

Write your answers below or on another piece of paper.

Find the perimeter (P) of each figure.

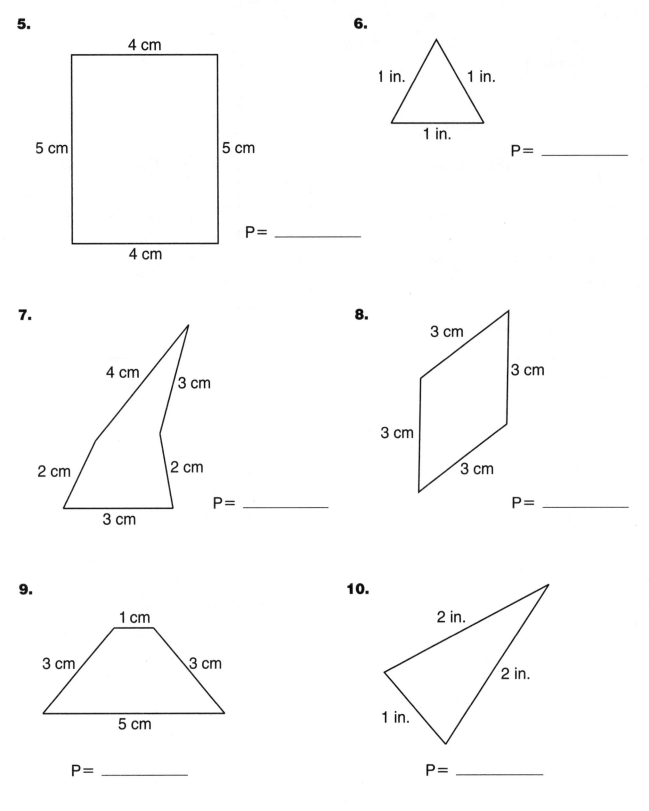

5.

4 cm

5 cm   5 cm

4 cm

P= _____

6.

1 in.   1 in.

1 in.

P= _____

7.

4 cm

3 cm

2 cm   2 cm

3 cm

P= _____

8.

3 cm

3 cm

3 cm

3 cm

P= _____

9.

1 cm

3 cm   3 cm

5 cm

P= _____

10.

2 in.

2 in.

1 in.

P= _____

Practice Set 20

**Use with or after
Lesson 3·8**

SRB
33
152 153

Write your answers below or on another piece of paper.

The diameter is given. Find the circumference (C) by using the "about 3 times" Circle Rule.

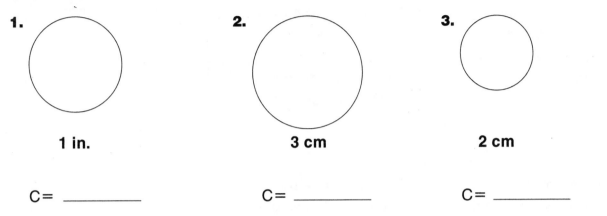

1.

1 in.

2.

3 cm

3.

2 cm

C= _____   C= _____   C= _____

Count the bills and coins. Then write the correct amount using dollars-and-cents notation.

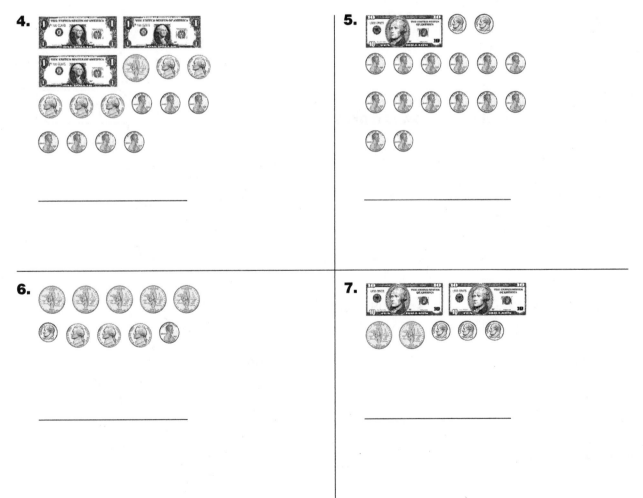

4.

_____

5.

_____

6.

_____

7.

_____

Practice Set 21

Write your answers below or on another piece of paper.

For Problems 1–4, use the diagram to help you find each answer.

1. David has 5 vases. He put 6 flowers in each vase. How many flowers did David put in all of the vases?

2. Sharon bought 4 packs of crackers. Each pack holds 8 crackers. How many crackers did Sharon buy in all?

Units		
Numbers		

_____

_____

3. A meeting room has 5 rows of chairs. Each row has 8 chairs. How many chairs are in all of the rows?

4. Each page of a photo album has 4 rows of pictures. Each row has 3 pictures. How many pictures are on each page of the album?

_____

_____

For Problems 5–17, add or subtract. Then make a ballpark estimate to check that your answer makes sense.

Unit
snowballs

5. $112 + 65 =$ _____

6. $197 - 53 =$ _____

7. $116 + 239 =$ _____

8. $456 - 327 =$ _____

9. $272 + 351 =$ _____

10. $923 - 685 =$ _____

11. $49 + 327 + 22 =$ _____

12. $708 - 349 =$ _____

13. $203 + 75 + 81 =$ _____

14.
$$\begin{array}{r} 152 \\ + 398 \\ \hline \end{array}$$

15.
$$\begin{array}{r} 941 \\ - 621 \\ \hline \end{array}$$

16.
$$\begin{array}{r} 384 \\ - 139 \\ \hline \end{array}$$

17.
$$\begin{array}{r} 516 \\ 225 \\ + 394 \\ \hline \end{array}$$

Practice Set 22

Use with or after
Lesson 4·2

Write your answers below or on another piece of paper.

For Problems 1–5, draw or build an array to help you solve each problem.

1. Tyler bought 3 boxes of snacks. Each box had 10 bags of snacks. How many bags of snacks did Tyler buy in all?

2. 4 pies are cut into 6 pieces each. How many pieces of pie are there in all?

3. Sharon bought 3 packs of invitations. Each pack had 8 invitations. How many invitations did Sharon buy in all?

4. Nancy displays her glass animals in a case with 5 shelves. Nancy puts 4 animals on each shelf. How many animals are in her display case?

5. Steve needs 5 inches of ribbon for each puppet that he is making. How many inches of ribbon will he need for 8 puppets?

For Problems 6–9, find the total value of each set of money.

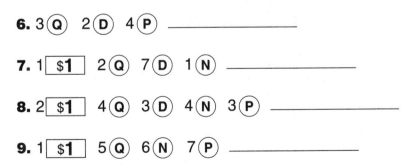

6. 3(Q) 2(D) 4(P) _____

7. 1 $1 2(Q) 7(D) 1(N) _____

8. 2 $1 4(Q) 3(D) 4(N) 3(P) _____

9. 1 $1 5(Q) 6(N) 7(P) _____

33

Practice Set 22 *continued*

Use with or after
Lesson 4·2

SRB 154–156

Write your answers below or on another piece of paper.

Find the area (A) of each rectangle or square in square units.

10.

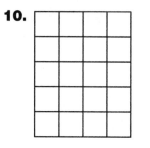

A = _____

11.

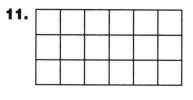

A = _____

12.

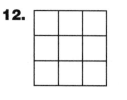

A = _____

13.

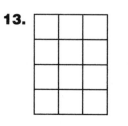

A = _____

14.

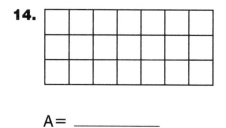

A = _____

15.

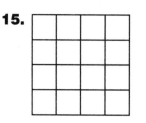

A = _____

Practice Set 23

Use with or after
Lesson 4·3

SRB
62
73 79

Write your answers below or on another piece of paper.

For Problems 1–6, use counters or draw pictures to help you solve the division problems.

24 grapes shared equally ...

1. *by 3 people*

2. *by 4 people*

3. *by 6 people*

_____ grapes
per person

_____ grapes
per person

_____ grapes
per person

_____ grapes
left over

_____ grapes
left over

_____ grapes
left over

48 cherries shared equally ...

4. *by 6 people*

5. *by 8 people*

6. *by 12 people*

_____ cherries
per person

_____ cherries
per person

_____ cherries
per person

_____ cherries
left over

_____ cherries
left over

_____ cherries
left over

Subtract.

Unit
crackerjacks

7. $89 - 67 =$ _____

8. $58 - 23 =$ _____

9. $90 - 36 =$ _____

10. $32 - 18 =$ _____

11. $77 - 56 =$ _____

12. $46 - 21 =$ _____

13.
$$\begin{array}{r} 73 \\ -\ 56 \\ \hline \end{array}$$

14.
$$\begin{array}{r} 53 \\ -\ 29 \\ \hline \end{array}$$

15.
$$\begin{array}{r} 62 \\ -\ 26 \\ \hline \end{array}$$

16.
$$\begin{array}{r} 63 \\ -\ 48 \\ \hline \end{array}$$

Practice Set 24

Use with or after
Lesson 4·4

SRB
12 73

Write your answers below or on another piece of paper.

For each problem, write a number model. Then find the missing numbers.

> **Example**   31 apples are divided evenly among 6 baskets.
> How many apples are in each basket?
>
> **Number model: 31 ÷ 6 → 5 R1**
>
> **5 apples are in each basket.**
> **1 apple is left over.**

1. 24 bones are shared
equally among 6 dogs.
How many bones
does each dog get?

_____ ÷ _____ → _____ R _____

Each dog gets _____ bones.

_____ bones are left over.

2. Tim has 27 jars of jam.
He puts 4 jars in each
box. How many boxes
does he fill?

_____ ÷ _____ → _____ R _____

Tim fills _____ boxes.

_____ jars are left over.

Find the missing numbers.

3.
4,010   4,040   4,050

4.
3,712   3,812   4,112

5.
2,115   3,115   7,115

Practice Set 24 *continued*

Use with or after
Lesson 4·4

SRB
203 204

Write your answers below or on another piece of paper.

Find the missing numbers.

6.

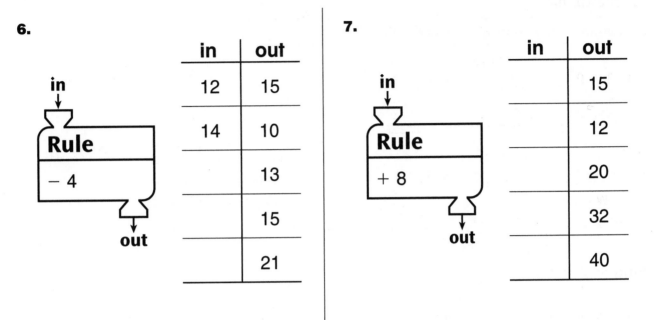

in	out
12	15
14	10
	13
	15
	21

7.

in	out
	15
	12
	20
	32
	40

Rule + 8

8.

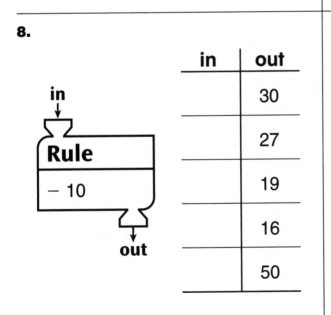

Rule − 10

in	out
	30
	27
	19
	16
	50

9.

Rule Double

in	out
	8
	14
	20
	24
	50

37

Practice Set 25

Write your answers below or on another piece of paper.

Solve each multiplication problem. Then write a turn-around shortcut for each problem.

Example	$6 \times 2 = 12$	$2 \times 6 = 12$

1. $3 \times 2 = $ _____

2. $4 \times 3 = $ _____

3. $3 \times 5 = $ _____

_____   _____   _____

4. $6 \times 3 = $ _____

5. $7 \times 4 = $ _____

6. $2 \times 4 = $ _____

_____   _____   _____

Find each product.

7. $5 \times 0 = $ _____

8. $7 \times 1 = $ _____

9. $3 \times 1 = $ _____

10. $16 \times 1 = $ _____

11. $0 \times 4 = $ _____

12. $12 \times 0 = $ _____

Write each amount with dollars and cents.

Example	$1 $1 Q D N N	$2.45

13. $1 Q Q D D P P P _____

14. $1 $1 $1 Q D D N P P _____

15. $10 $1 D D D D N P _____

16. $1 Q Q Q Q Q D P P P P _____

Practice Set 26

Write your answers below or on another piece of paper.

Find the missing number for each Fact Triangle. Then write the family of facts for that triangle.

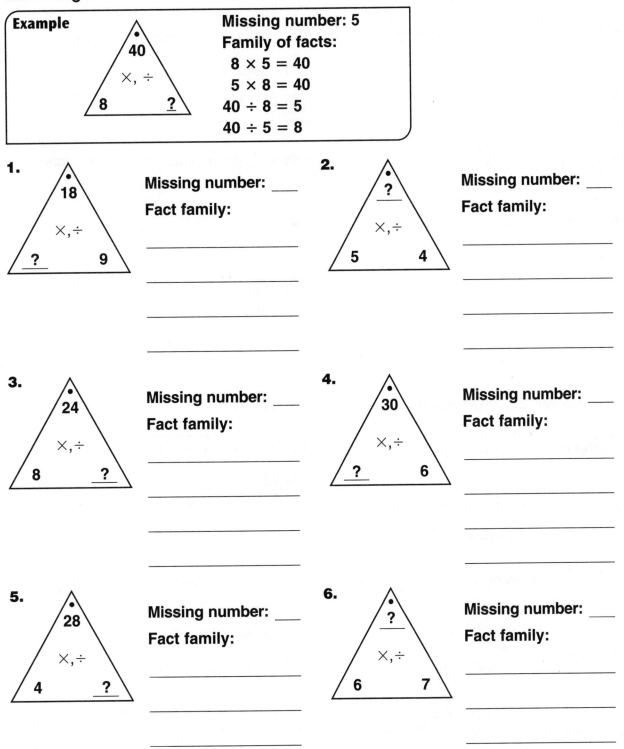

Example

40
×, ÷
8   ?

Missing number: 5
Family of facts:
    8 × 5 = 40
    5 × 8 = 40
    40 ÷ 8 = 5
    40 ÷ 5 = 8

1.

18
×, ÷
?   9

Missing number: ___
Fact family:

_____

_____

_____

2.

?
×, ÷
5   4

Missing number: ___
Fact family:

_____

_____

_____

3.

24
×, ÷
8   ?

Missing number: ___
Fact family:

_____

_____

_____

4.

30
×, ÷
?   6

Missing number: ___
Fact family:

_____

_____

_____

5.

28
×, ÷
4   ?

Missing number: ___
Fact family:

_____

_____

_____

6.

?
×, ÷
6   7

Missing number: ___
Fact family:

_____

_____

_____

39

Practice Set 26 *continued*

Write your answers below or on another piece of paper.

Match each amount of money with an equal amount from the list at right. Then write the letter that identifies that amount.

7. fourteen dollars and two cents _____

A. twelve dollars and forty cents

8. $20.14 _____

B. $12.04

9. $41.20 _____

C. forty-one dollars and twenty cents

10. $12.40 _____

D. $1.42

11. $10.42 _____

E. twenty dollars and fourteen cents

12. twelve dollars and four cents _____

F. ten dollars and forty-two cents

13. one dollar and forty-two cents _____

G. $14.02

Find the missing numbers. You can use counters or draw pictures.

14. 15 pieces of candy
4 children share equally

_____ pieces per child

_____ pieces remaining

15. 12 tennis balls
3 balls per can

_____ filled cans

_____ balls remaining

16. 14 carrots
6 rabbits share equally

_____ carrots per rabbit

_____ carrots remaining

17. 27 books
8 books per box

_____ filled boxes

_____ books remaining

40

Practice Set 27

Use with or after
Lesson 4·8

SRB
18–20
52–53

Write your answers below or on another piece of paper.

Multiply the number in the *center* of the circle by each number *on* the circle. Then write the product *outside* the circle.

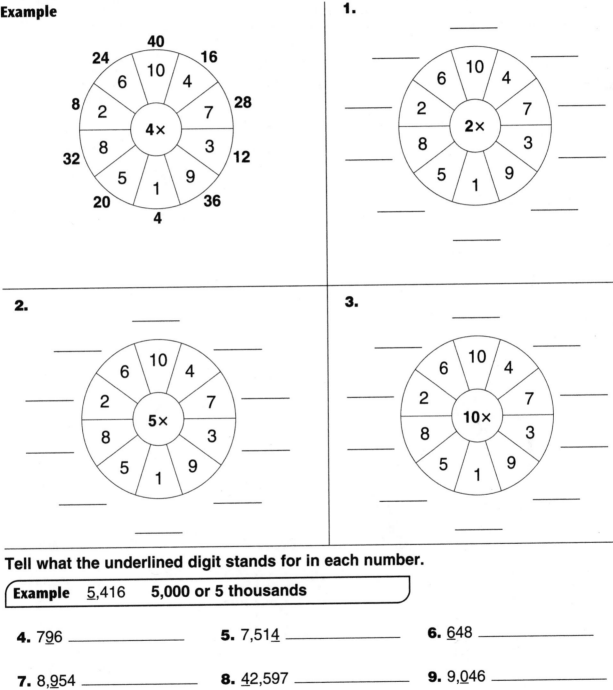

Example

1.

2.

3.

Tell what the underlined digit stands for in each number.

Example 5,416   5,000 or 5 thousands

4. 7<u>9</u>6 _____

5. 7,51<u>4</u> _____

6. <u>6</u>48 _____

7. 8,<u>9</u>54 _____

8. <u>4</u>2,597 _____

9. 9,<u>0</u>46 _____

10. 58,0<u>2</u>9 _____

11. <u>6</u>,309 _____

12. 3,<u>1</u>89 _____

Practice Set 27 *continued*

Write your answers below or on another piece of paper.

Write a multiplication fact to find the total number of dots in each array.

Example		
	• •	$3 \times 7 = 21$ 21 total dots

13. • • • •
        • • • •
        • • • •
        • • • •

_____

_____

14. • • • •
        • • • •
        • • • •
        • • • •
        • • • •

_____

_____

15. • •
        • •
        • •
        • •
        • •

_____

_____

16. • • •
        • • •
        • • •
        • • •
        • • •

_____

_____

17. • • • • • • • •
        • • • • • • • •
        • • • • • • • •

_____

_____

18. • • • • •
        • • • • •
        • • • • •
        • • • • •

_____

_____

Name _____ Date _____ Time _____

Practice Set 28

**Use with or after
Lesson 5·1**

SRB
18–21
52 53

Write your answers below or on another piece of paper.

Answer each question using the numbers in the box.

| 79,512 | 29,517 | 12,759 | 27,951 | 95,721 |

1. Which numbers have 5 tens? _____

2. Which number has 5 thousands? _____

3. Which numbers have 1 ten? _____

4. Which numbers have 9 thousands? _____

5. Which number has 9 ten-thousands? _____

6. Which numbers have 7 hundreds? _____

7. Which number has 2 ones? _____

8. Which numbers have 5 hundreds? _____

9. Which numbers have 2 ten-thousands? _____

10. Which numbers have 1 one? _____

11. Which number has 7 thousands? _____

Find each product.

12. $1 \times 7 =$ _____

13. $5 \times 2 =$ _____

14. $8 \times 2 =$ _____

15. $3 \times 10 =$ _____

16. $6 \times 5 =$ _____

17. $9 \times 5 =$ _____

18. $10 \times 5 =$ _____

19. $8 \times 0 =$ _____

20. $6 \times 1 =$ _____

Unit

pink elephants

43

Practice Set 28 *continued*

Use with or after
Lesson 5·1

SRB
200 201

Write your answers below or on another piece of paper.

Find the missing rule and the numbers for the empty frames.

Example

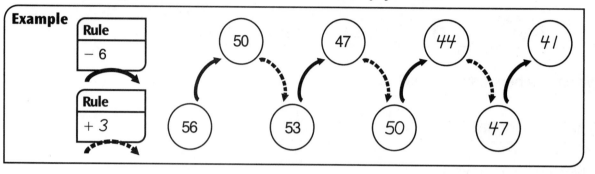

21.

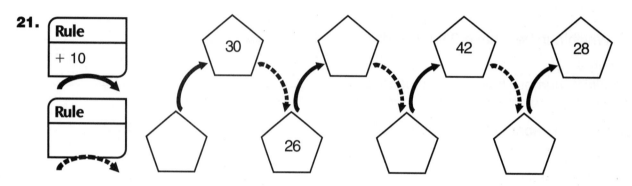

22.

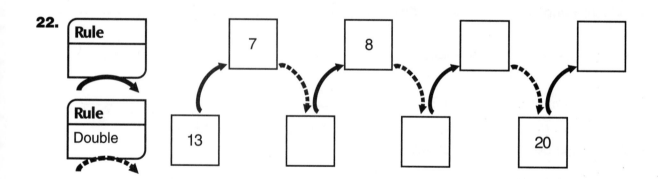

23.

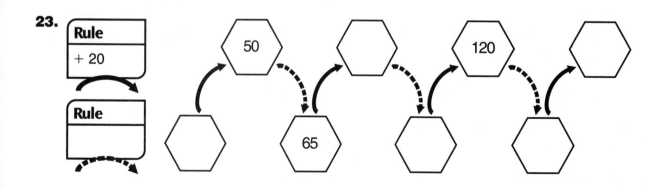

Practice Set 29

Write your answers below or on another piece of paper.

Write each group of numbers in order from smallest to largest.

1. 4,289   4,892   2,489   9,842   4,982

_____

2. 5,901   6,001   5,991   1,995   5,910   6,010

_____

3. 10,453   1,543   11,246   21,101   9,878

_____

4. 71,034   17,340   80,249   99,999   81,001

_____

Use your calculator to count by 100s. Find the missing numbers.

5. 2,300; 2,400; _____; _____; _____; 2,800; _____; _____; 3,100; 3,200

6. 7,260; 7,360; _____; _____; _____; 7,760; _____; _____; 8,060; _____

7. 5,408; 5,508; _____; _____; 5,808; 5,908; _____; _____; _____

8. 6,132; _____; _____; _____; 6,532; _____; _____; 6,832; _____; _____

9. 8,555; 8,655; _____; _____; _____; 9,055; _____; _____; _____; _____

Find the same time in the second list. Write the letter that identifies that matching time.

10. 5 minutes after 3 _____   **A.** 5:30

11. 10:40 _____   **B.** half-past 7

12. quarter-to 5 _____   **C.** 3:05

13. 7:30 _____   **D.** 20 minutes to 11

14. five-thirty _____   **E.** 10:15

15. quarter-after 10 _____   **F.** 4:45

Practice Set 30

Write your answers below or on another piece of paper.

Write the following numbers using digits:

1. one million, two hundred twenty-eight thousand _____

2. six million, three thousand, four hundred _____

3. seven hundred thirty-one thousand, five hundred forty-nine _____

4. eighty-three thousand, nine hundred two _____

5. four million, six hundred five _____

6. three million, twenty thousand, five hundred _____

Add or subtract. Then make a ballpark estimate to check that your answer makes sense.

7. $721 - 350 =$ _____   **8.** $213 + 643 =$ _____   **9.** $672 - 514 =$ _____

10. $815 + 192 =$ _____   **11.** $728 - 456 =$ _____   **12.** $359 + 287 =$ _____

13. 821
      $- 371$

14. 416
      $- 203$

15. 89
      $+ 376$

16. 327
      $- 119$

17. 223
      $+ 478$

18. 632
      $- 218$

19. 551
       321
      $+ 114$

20. 264
       118
      $+ 319$

Practice Set 31

Use with or after
Lesson 5·4

SRB
13

Write your answers below or on another piece of paper.

Write < or >.

Example	59,423 _____ 59,389

Both numbers have 5 ten-thousands and 9 thousands.

4 hundreds is greater than 3 hundreds.

Therefore: **59,423 > 59,389**

< means *is less than*
> means *is greater than*

1. 127,675 _____ 137,675

2. 24,714 _____ 24,710

3. 159,338 _____ 160,273

4. 673,218 _____ 673,239

5. 285,641 _____ 385,641

6. 490,315 _____ 510,214

7. 331,846 _____ 330,259

8. 37,014 _____ 37,104

9. 999,972 _____ 999,992

10. 53,892 _____ 54,617

For each number below, write the number that is 10 more, 100 more, and 1,000 more.

Example	33,492

10 more: 33,502
100 more: 33,592
1,000 more: 34,492

11. 52,416

12. 68,927

13. 72,499

_____
_____
_____

_____
_____
_____

_____
_____
_____

14. 65,798

15. 95,281

16. 39,482

_____
_____
_____

_____
_____
_____

_____
_____
_____

47

Practice Set 31 *continued*

Use with or after
Lesson 5·4

SRB
52 53
140

Write your answers below or on another piece of paper.

Find each missing number.

> 1 meter = 10 decimeters
> 1 meter = 100 centimeters
> 1 decimeter = 10 centimeters

17. 3 meters = _____ decimeters

18. 300 centimeters = _____ meters

19. 70 centimeters = _____ decimeters

20. 8 meters = _____ decimeters

21. 4 decimeters = _____ centimeters

22. _____ centimeters = 5 meters

23. 400 centimeters = _____ meters

24. 20 decimeters = _____ meters

25. 100 decimeters = _____ meters

26. _____ decimeters = 100 centimeters

Find each answer.

27. $2 \times 6 =$ _____   **28.** $1 \times 18 =$ _____   **29.** $5 \times 4 =$ _____

30. $7 \times 3 =$ _____   **31.** $8 \times 5 =$ _____   **32.** $6 \times 5 =$ _____

33. $\begin{array}{r} 9 \\ \times\, 6 \\ \hline \end{array}$ **34.** $\begin{array}{r} 6 \\ \times\, 7 \\ \hline \end{array}$ **35.** $\begin{array}{r} 3 \\ \times\, 8 \\ \hline \end{array}$ **36.** $\begin{array}{r} 4 \\ \times\, 2 \\ \hline \end{array}$

37. $\begin{array}{r} 0 \\ \times\, 29 \\ \hline \end{array}$ **38.** $\begin{array}{r} 5 \\ \times\, 9 \\ \hline \end{array}$ **39.** $\begin{array}{r} 2 \\ \times\, 10 \\ \hline \end{array}$ **40.** $\begin{array}{r} 7 \\ \times\, 3 \\ \hline \end{array}$

Practice Set 32

Write your answers below or on another piece of paper.

Write a decimal for the shaded part of each grid. Each grid is ONE.

Example

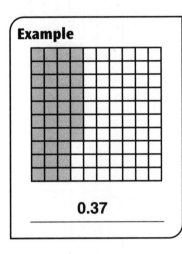

0.37

1.

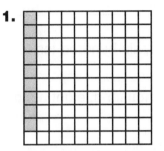

_____

2.

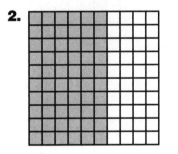

_____

3.

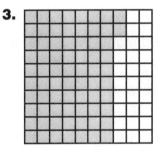

_____

4.

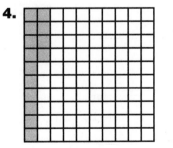

_____

5.

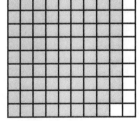

_____

49

Practice Set 32 continued

Write your answers below or on another piece of paper.

Write each number using digits.

> **Example**   twenty-six thousand, four hundred twelve
> **26,412**

6. four thousand, six hundred ten _____

7. seventy-two thousand, eight hundred five _____

8. five thousand, six _____

9. sixty-seven thousand, three hundred eighteen _____

10. one hundred fourteen thousand, five hundred thirty-six _____

11. seven hundred eighty thousand, two hundred _____

12. six hundred seventy-nine thousand, eight _____

13. four hundred three thousand, ninety-two _____

Use your calculator.

Write the answer in dollars and cents.

> **Example** $2.71 + 92¢ = **$3.63**

14. 67¢ + $1.21 = _____

15. $0.59 + 83¢ + 27¢ = _____

16. $3.41 + $1.08 = _____

17. 114¢ + 32¢ = _____

18. $0.65 + $0.84 = _____

19. $5.04 + 42¢ + 33¢ = _____

20. $0.84 + $2.86 = _____

21. 256¢ + $2.56 = _____

22. 49¢ + $1.92 + 67¢ = _____

23. $4.41 + $3.15 + $6.85 = _____

Practice Set 33

Write your answers below or on another piece of paper.

Use the decimals in the box to answer each question below.

| 0.03 | 0.34 | 0.45 | 0.56 | 0.67 | 0.78 | 0.89 |

1. Which number has 3 tenths? _____

2. Which number has 5 hundredths? _____

3. Which number has 7 tenths? _____

4. Which number has 9 hundredths? _____

5. Which number has 0 tenths? _____

6. Which number has 6 tenths and 7 hundredths? _____

7. Which number has 5 tenths and 6 hundredths? _____

8. Which number has 3 tenths and 4 hundredths? _____

Write the addition and subtraction fact family for each group of numbers.

Example   5, 6, 11
$$5 + 6 = 11$$
$$6 + 5 = 11$$
$$11 - 6 = 5$$
$$11 - 5 = 6$$

9. 2, 8, 10   **10.** 7, 8, 15   **11.** 9, 3, 12

51

Practice Set 33 *continued*

Use with or after
Lesson 5·8

SRB
182–185

Write your answers below or on another piece of paper.

For each number, write the number that is 10 less, 100 less, and 1,000 less.

Example   52,928
            10 less: 52,918
            100 less: 52,828
            1,000 less: 51,928

12. 27,386   **13.** 50,221   **14.** 38,482

_____   _____   _____

_____   _____   _____

_____   _____   _____

15. 93,525   **16.** 27,058   **17.** 60,867

_____   _____   _____

_____   _____   _____

_____   _____   _____

18. 62,505   **19.** 28,331   **20.** 83,471

_____   _____   _____

_____   _____   _____

_____   _____   _____

Solve each problem. You can use counters or draw pictures.

21. David had 527 baseball cards in his collection. He got 85 more baseball cards for his birthday. How many baseball cards does he have now?

22. Jenny has 412 hockey cards and 843 basketball cards. How many more basketball cards than hockey cards does she have?

Practice Set 34

Use with or after
Lesson 5·9

SRB
13
52–53

Write your answers below or on another piece of paper.

Write <, >, or =.

| = means *is equal to* |
| < means *is less than* |
| > means *is greater than* |

1. 0.01 meter _____ 0.08 meter

2. 0.30 meter _____ 0.03 meter

3. 0.40 meter _____ 0.20 meter

4. 0.50 meter _____ 0.54 meter

5. 0.1 meter _____ 2 longs

6. 0.01 meter _____ 5 cubes

7. 0.07 meter _____ 7 cubes

8. 0.6 meter _____ 6 cubes

■ ← cube (1 cm)

▬▬ ← long (10 cm)

← meterstick

Write the multiplication and division fact family for each group of numbers.

Example   16, 8, 2
$$8 \times 2 = 16$$
$$2 \times 8 = 16$$
$$16 \div 2 = 8$$
$$16 \div 8 = 2$$

9. 4, 5, 20

10. 2, 4, 8

11. 3, 4, 12

12. 5, 2, 10

13. 9, 2, 18

14. 5, 6, 30

53

Practice Set 35

Write your answers below or on another piece of paper.

Write all of your answers.

```
|||||||||||||||||||||||||||||||||||||||||||||||||||||||||||||||||||||||||||||||
 0   1   2   3   4   5   6   7   8   9   10  11  12  13  14  15
cm
```

1. How many millimeters in a centimeter? _____

2. How many centimeters in a meter? _____

3. How many millimeters in 2 centimeters? _____

4. How many millimeters in a meter? _____

5. How many millimeters in 24.3 centimeters? _____

Estimate to answer *yes* or *no* to each question.

6. You have $9.00.
Do you have enough
to buy a book for $3.60,
a magazine for $2.29,
and a poster for $2.79?

_____

7. You have $12.00.
Do you have enough
to buy a beach towel
for $7.30 and
sunglasses for $4.45?

_____

8. You have $20.00.
Do you have enough
to buy a shirt
for $11.40 and
shorts for $7.39?

_____

9. You have $15.00.
Do you have enough
to buy a fishing pole
for $9.50, fishing line
for $3.89, and bait
for $1.79?

_____

Practice Set 35 *continued*

Use with or after
Lesson 5·10

SRB
12
140

Write your answers below or on another piece of paper.

Find each missing number.

1 m = 10 dm	1 dm = 0.1 m
1 m = 100 cm	1 cm = 0.01 m
1 dm = 10 cm	1 cm = 0.1 dm

10. _____ m = 400 cm

11. 0.8 dm = _____ cm

12. 60 cm = _____ dm

13. 0.5 m = _____ dm

14. _____ dm = 9 m

15. _____ cm = 0.02 m

16. 7 m = _____ cm

17. 0.4 m = _____ dm

18. 8 cm = _____ m

19. _____ m = 9 cm

20. _____ cm = 3 m

21. _____ cm = 0.9 dm

Find the missing numbers.

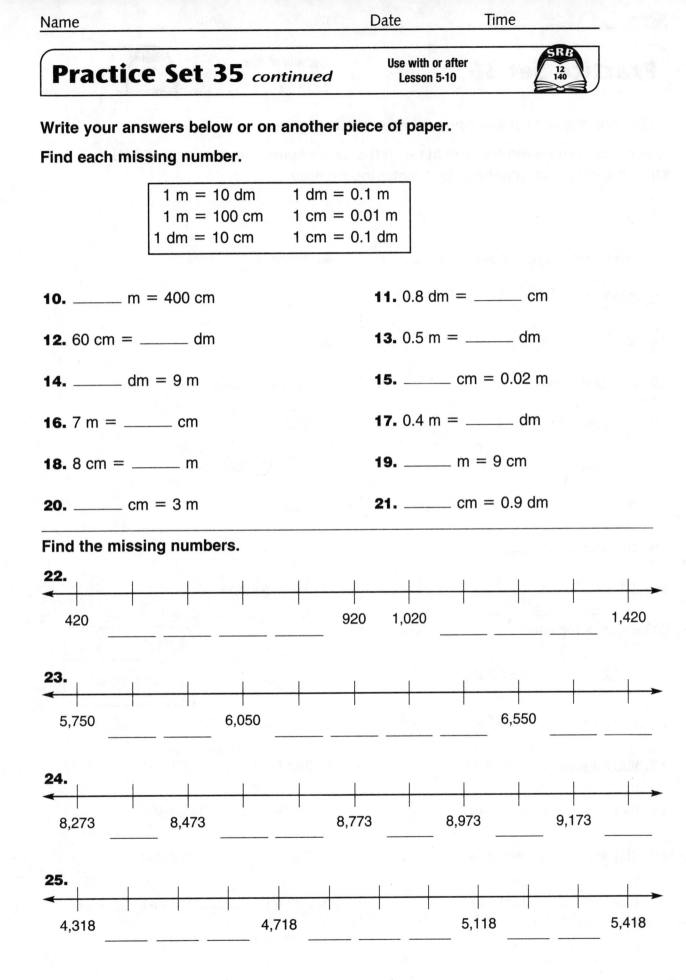

22.

420   920   1,020   1,420

23.

5,750   6,050   6,550

24.

8,273   8,473   8,773   8,973   9,173

25.

4,318   4,718   5,118   5,418

Practice Set 36

Use with or after
Lesson 5·11

SRB
13
35 36

Write your answers below or on another piece of paper.

Match each number in the first list with the same number in the second list. Write the letter that identifies that matching number.

1. .5 _____

A. .004

2. two hundred one thousandths _____

B. 2 tenths

3. .521 _____

C. .040

4. .05 _____

D. 5 tenths

5. 21 thousandths _____

E. 21 hundredths

6. 4 thousandths _____

F. .201

7. .4 _____

G. 521 thousandths

8. .2 _____

H. .021

9. 40 thousandths _____

I. 4 tenths

10. .21 _____

J. 5 hundredths

Write < or > for each.

Unit
harbor seals

11. 465,243 _____ 564,243

12. 107,453 _____ 107,452

13. 999,999 _____ 1,000,000   14. 382,591 _____ 382,491

15. 848,484 _____ 484,848   16. 12,495 _____ 112,495

17. 359,416 _____ 369,416   18. 992,450 _____ 992,460

19. 600,770 _____ 600,707   20. 739,418 _____ 843,291

Practice Set 37

Write your answers below or on another piece of paper.

Count the number of line segments used to make each figure.

1.

_____

2.

_____

3.

_____

4.

_____

For each problem below, write a number model.
Then find the missing numbers.

5. 12 pens are shared equally among
4 children. How many pens does
each child get?

_____ ÷ _____ → _____ R _____

Each child gets _____ pens.

_____ pens are left over.

6. 8 toy mice are shared equally among
3 cats. How many mice does
each cat get?

_____ ÷ _____ → _____ R _____

Each cat gets _____ mice.

_____ mice are left over.

7. Brian has 11 sweaters and puts 3 in each
drawer. How many drawers does Brian fill?

_____ ÷ _____ → _____ R _____

Brian fills _____ drawers.

_____ sweaters are left over.

Practice Set 38

Use with or after
Lesson 6·2

SRB
52–53
99–100

Write your answers below or on another piece of paper.

Match each description with the correct example. Write the letter that identifies that example.

1. parallel lines _____

A.

2. intersecting lines _____

B.

3. intersecting line segments _____

C.

4. parallel rays _____

D.

5. Draw a pair of parallel line segments.

6. Draw a pair of intersecting rays.

Write the multiplication and division fact family for each group of numbers.

7. 25, 5, 5

_____

_____

_____

8. 2, 4, 2

_____

_____

_____

9. 8, 64, 8

_____

_____

_____

10. 9, 9, 81

_____

_____

_____

11. 42, 6, 7

_____

_____

_____

12. 7, 7, 49

_____

_____

_____

13. 8, 72, 9

_____

_____

_____

14. 6, 30, 5

_____

_____

_____

15. 27, 9, 3

_____

_____

_____

Practice Set 38 *continued*

Use with or after
Lesson 6·2

SRB
134–135
143–145

Write your answers below or on another piece of paper.

Measure each object to the nearest half-inch or half-centimeter.

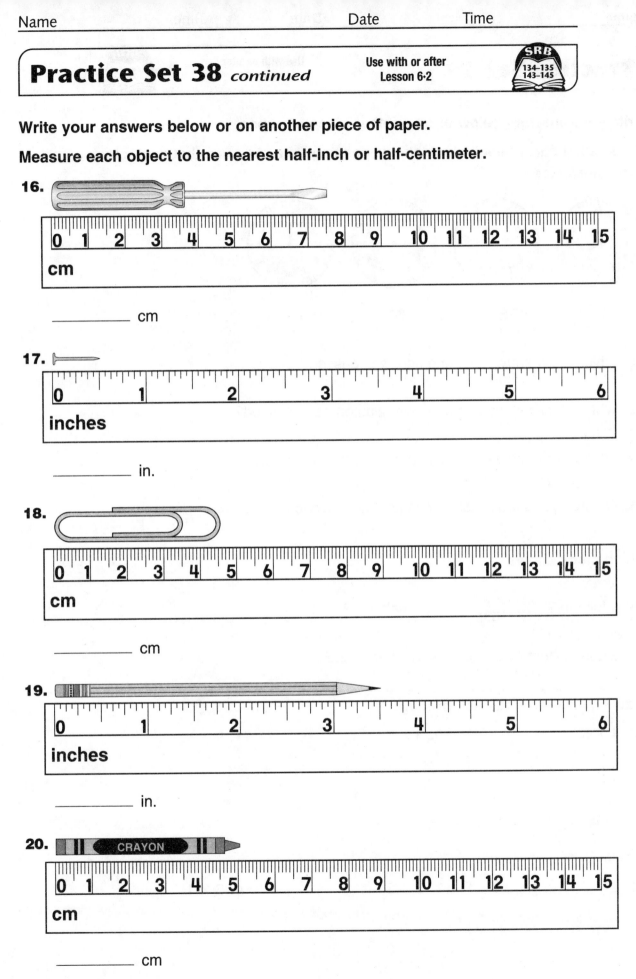

16.

_____ cm

17.

_____ in.

18.

_____ cm

19.

_____ in.

20.

_____ cm

Name _____   Date _____   Time _____

Practice Set 39

Use with or after
Lesson 6·3

SRB
64–67
174

Write your answers below or on another piece of paper.

The shaded part of each clock shows passing time. Assume that each clock turns clockwise.

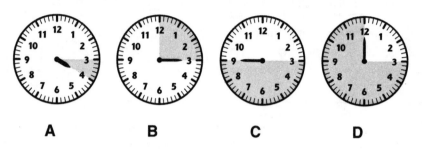

A   B   C   D

1. Which clock shows that one hour has passed? _____

2. Which clock shows that forty-five minutes have passed? _____

3. Which clock shows that half an hour has passed? _____

4. Which clock shows that 15 minutes have passed? _____

Which clock shows ...

5. a full turn? _____   **6.** a half-turn? _____

7. a quarter-turn? _____   **8.** a $\frac{3}{4}$ turn? _____

Draw an array to find each product.

9. 4×5 **10.** 3×8 **11.** 9×4

12. 2×7 **13.** 5×5 **14.** 6×1

60

Practice Set 39 *continued*

Use with or after
Lesson 6·3

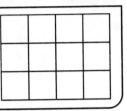

Write your answers below or on another piece of paper.

Find the area of each rectangle or square in square centimeters. Find the perimeter of each rectangle or square in centimeters.

Example

Area: 12 square centimeters
Perimeter: 14 centimeters

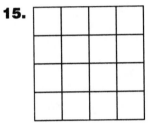

15.

Area: _____

Perimeter: _____

16.

Area: _____

Perimeter: _____

17.

Area: _____

Perimeter: _____

18.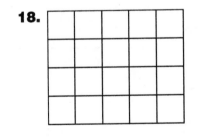

Area: _____

Perimeter: _____

Practice Set 40

Use with or after
Lesson 6·6

SRB
18–20
98–104

Write your answers below or on another piece of paper.

Write all of your answers below or on a separate piece of paper.

1. Circle the shapes that have right angles.

A.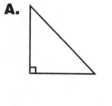

B.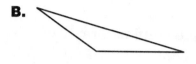

C.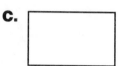

D.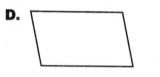

E.

Write the number that has ...

Example	4 tens	**30,947**
	9 hundreds	
	0 thousands	
	3 ten-thousands	
	7 ones	

2. 8 hundreds _____
   9 ones
   5 ten-thousands
   3 tens
   9 thousands

3. 4 thousands _____
   6 tens
   1 hundred
   2 ones
   7 ten-thousands

4. 7 hundreds _____
   0 ones
   3 ten-thousands
   4 thousands
   9 tens

5. 8 thousands _____
   5 tens
   2 hundreds
   6 ten-thousands
   8 ones

Practice Set 40 *continued*

Use with or after
Lesson 6·6

SRB
150–151

Write your answers below or on another piece of paper.

Match each description with the correct polygon. Write the letter of that polygon.

6. a rectangle with a perimeter of 22 in. _____

7. a triangle with a perimeter of 18 in. _____

8. a parallelogram with a perimeter of 18 in. _____

9. a square with a perimeter of 16 in. _____

10. a kite with a perimeter of 18 in. _____

11. a triangle with a perimeter of 17 in. _____

12. a rhombus with a perimeter of 28 in. _____

13. a rectangle with a perimeter of 20 in. _____

A. 4 in.

B. 5 in. 5 in. 7 in.

C. 4 in. 7 in.

D. 6 in. 4 in. 8 in.
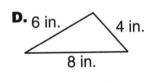

E. 4 in. 5 in.
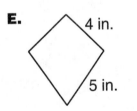

F. 7 in.
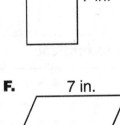

G. 4 in. 6 in.
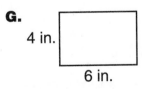

H. 7 in. 2 in.
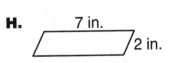

63

Practice Set 41

Use with or after
Lesson 6·7

SRB
64–67
98

Write your answers below or on another piece of paper.

Draw each angle as directed below. Record the direction of each turn with a curved arrow. And mark with a □ any right angle you make.

Example   Angle *H* shows a $\frac{3}{4}$ turn

1. An angle that shows a quarter-turn

2. An angle that shows a half-turn

3. An angle that shows a $\frac{3}{4}$ turn

4. An angle that is smaller than a half-turn

5. An angle that is larger than a half-turn

Draw an array to find each product.

Example 3×7 • • • • • • •   21
                        • • • • • • •
                        • • • • • • •

6. 2×5 **7.** 6×4 **8.** 1×8 **9.** 9×3

10. 5×6 **11.** 8×3 **12.** 5×1 **13.** 7×6

Practice Set 42

Use with or after
Lesson 6·9

SRB
122
123

Write your answers below or on another piece of paper.

Each picture below shows one-half of a letter. The dashed line is the line of symmetry. Write the complete letter.

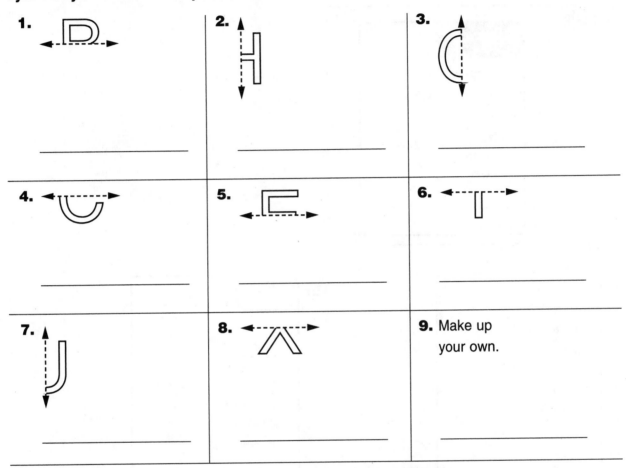

1. _____

2. _____

3. _____

4. _____

5. _____

6. _____

7. _____

8. _____

9. Make up
   your own.
   _____

Draw the number of lines of symmetry shown in parentheses.

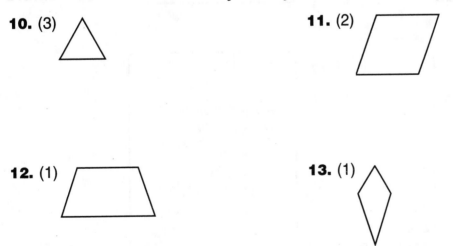

10. (3)

11. (2)

12. (1)

13. (1)

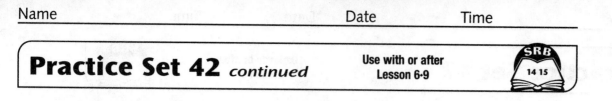

Practice Set 42 *continued*

Use with or after Lesson 6·9

SRB
14 15

Write your answers below or on another piece of paper.

Cross out the names that DO NOT belong in each name-collection box. Then write the number that belongs on the label for each box.

Example

2 1

5 + 8 + 5 + 6 ~~crossed out~~

16 + 5   30 − 12 ~~crossed out~~

10 + 5 + 6

~~18 − 2~~ ~~crossed out~~

twenty-one

~~8 × 3~~ ~~crossed out~~   10 + 11

14.

4 × 8   15
                + 15
3 × 10   7
            × 4
20 + 13

8 + 9 + 13

5 more than 25

5 × 6   3 × 9

15.

20 − 2   9
                   + 9
● ● ● ● ● ● ● ● ● ●
● ● ● ● ● ● ● ● ●
4 + 5 + 9   2 × 9

2 more than 15

卌 卌 卌 | 8
                  + 9
4 less than 23

16.

2 × 20   5
                 × 8
9 × 5

6 less than 45

10 + 10 + 10 + 10

40 × 0   forty-one

14 + 26

5 more than 35

17.

20 less than 60   25
                      + 25
14 + 36   10
               × 5
6 × 10

20 + 20 + 10

0 × 50

30 + 25   100 − 5

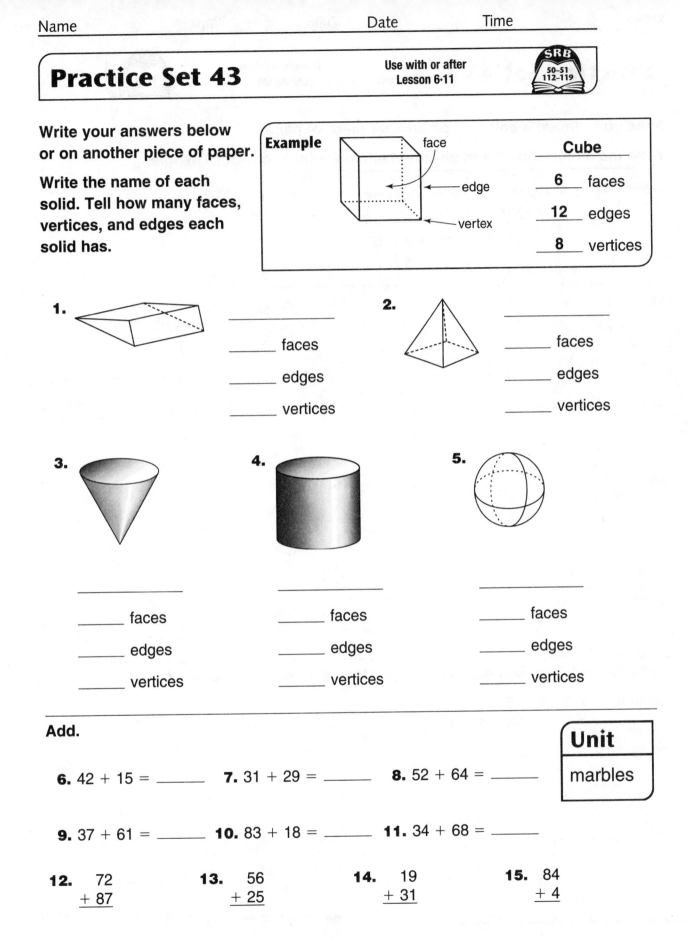

Practice Set 43

Write your answers below or on another piece of paper.

Write the name of each solid. Tell how many faces, vertices, and edges each solid has.

Example

face

edge

vertex

Cube

__6__ faces

__12__ edges

__8__ vertices

1. _____

_____ faces

_____ edges

_____ vertices

2. _____

_____ faces

_____ edges

_____ vertices

3. _____

_____ faces

_____ edges

_____ vertices

4. _____

_____ faces

_____ edges

_____ vertices

5. _____

_____ faces

_____ edges

_____ vertices

Add.

6. 42 + 15 = _____   **7.** 31 + 29 = _____   **8.** 52 + 64 = _____

9. 37 + 61 = _____   **10.** 83 + 18 = _____   **11.** 34 + 68 = _____

12. 72
       + 87

13. 56
       + 25

14. 19
       + 31

15. 84
       + 4

Unit

marbles

67

Name _____ Date _____ Time _____

Practice Set 43 *continued*

Use with or after
Lesson 6-11

SRB
52-55

Write your answers below or on another piece of paper.

Write the multiplication and division fact family for each Fact Triangle.

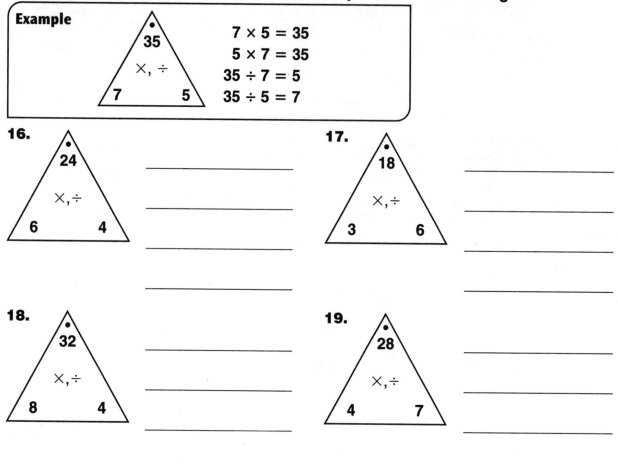

Example

35
×, ÷
7   5

7 × 5 = 35
5 × 7 = 35
35 ÷ 7 = 5
35 ÷ 5 = 7

16.
24
×, ÷
6   4

_____
_____
_____

17.
18
×, ÷
3   6

_____
_____
_____

18.
32
×, ÷
8   4

_____
_____
_____

19.
28
×, ÷
4   7

_____
_____
_____

Write the multiplication and division fact family for each group of numbers.

Example   5, 5, 25   5 × 5 = 25
                           25 ÷ 5 = 5

20. 3, 3, 9 _____

21. 16, 4, 4 _____

22. 6, 36, 6 _____

23. 49, 7, 7 _____

24. 9, 81, 9 _____

25. 8, 64, 8 _____

68

Practice Set 44

Use with or after
Lesson 6·12

SRB
115–117

Write your answers below or on another piece of paper.

Sandy is making a design by pressing the bases of pyramids and prisms onto an ink pad. What shape can she make from each block?

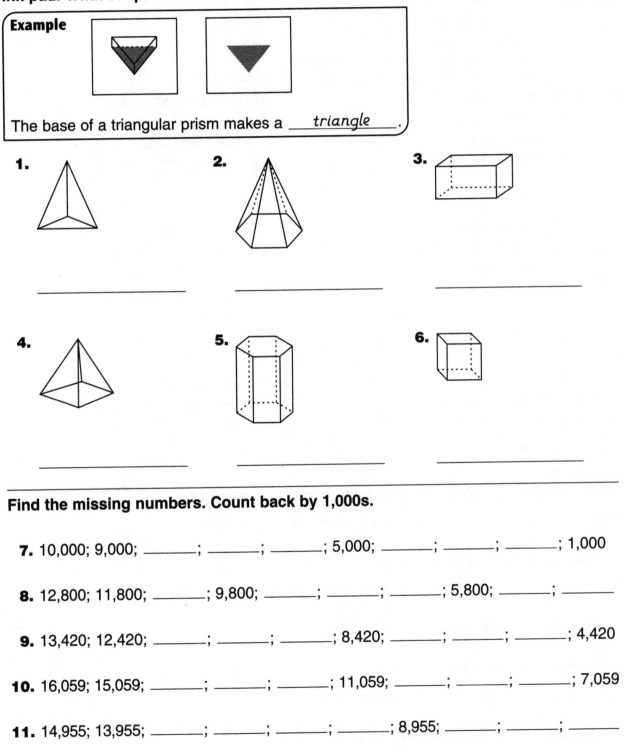

Example

The base of a triangular prism makes a ___*triangle*___.

1. _____

2. _____

3. _____

4. _____

5. _____

6. _____

Find the missing numbers. Count back by 1,000s.

7. 10,000; 9,000; _____; _____; _____; 5,000; _____; _____; _____; 1,000

8. 12,800; 11,800; _____; 9,800; _____; _____; _____; 5,800; _____; _____

9. 13,420; 12,420; _____; _____; _____; 8,420; _____; _____; _____; 4,420

10. 16,059; 15,059; _____; _____; _____; 11,059; _____; _____; _____; 7,059

11. 14,955; 13,955; _____; _____; _____; _____; 8,955; _____; _____; _____

Practice Set 44 *continued*

Use with or after
Lesson 6·12

SRB
18–20
52–53

Write your answers below or on another piece of paper.

Write the number that has ...

Example	2 in the tenths place
	9 in the ones place
	6 in the tens place
	8 in the hundredths place **69.28**

12. 0 in the ones place
4 in the tenths place
8 in the tens place
9 in the hundredths place

_____

13. 8 in the thousandths place
2 in the tenths place
6 in the ones place
5 in the hundredths place

_____

14. 7 in the tenths place
0 in the hundredths place
9 in the tens place
4 in the ones place

_____

15. 9 in the ones place
4 in the hundredths place
3 in the tenths place
8 in the thousandths place

_____

16. 2 in the tens place
5 in the tenths place
9 in the ones place
6 in the hundredths place

_____

17. 0 in the tenths place
4 in the thousandths place
7 in the hundredths place
2 in the ones place

_____

Solve each problem. You can draw pictures or use counters.

18. Ginny bought 4 boxes of markers. Each box has 8 markers.

How many markers did Ginny buy? _____

19. John has 18 roses. He puts 9 roses in each vase.

How many vases does he fill? _____

How many roses are left over? _____

20. Lia shared 22 cookies equally among 6 friends.

How many cookies did each friend get? _____

How many cookies were left over? _____

Practice Set 45

Use with or after
Lesson 7·1

SRB
64–67
73 137

Write your answers below or on another piece of paper.

Draw a square array for each square number. Then write the multiplication fact for each square number.

Example 25 •••••
•••••
•••••
•••••
••••• 5 × 5 = 25

1. 36

2. 16

3. 9

4. 4

5. 49

6. 64

Find the missing numbers. You can use counters or draw pictures.

7. 26 crackers
6 children share equally

_____ crackers per child

_____ crackers left over

8. 36 pictures
9 pictures per page

_____ filled pages

_____ pages left over

9. 24 girls
4 girls per tent

_____ filled tents

_____ girls left over

10. 18 sheets of paper
4 children share equally

_____ sheets per child

_____ sheets left over

Find the corresponding letter on the centimeter ruler for each of the metric measures.

Example 2.5 centimeters is Point *A.*

```
   A G   D      B   E    F    C
   • •   •      •   •    •    •
```

0 1 2 3 4 5 6 7 8 9 10 11 12 13 14 15
cm

11. 60 millimeters _____

12. 1 decimeter _____

13. 0.04 meter _____

14. 0.7 decimeter _____

15. 85 millimeters _____

16. 3 centimeters _____

Practice Set 45 *continued*

Use with or after
Lesson 7·1

SRB
50–51

Write your answers below or on another piece of paper.

Write the number that is 10 more.

17. 14 **18.** 30 **19.** 539 **20.** 4,258 **21.** 7,904

_____ _____ _____ _____ _____

Write the number that is 100 more.

22. 8 **23.** 27 **24.** 973 **25.** 2,918 **26.** 8,715

_____ _____ _____ _____ _____

Write the number that is 1,000 more.

27. 7 **28.** 254 **29.** 5,791 **30.** 9,493 **31.** 12,463

_____ _____ _____ _____ _____

Write the number that is 10 less.

32. 19 **33.** 142 **34.** 1,014 **35.** 7,420 **36.** 4,615

_____ _____ _____ _____ _____

Write the number that is 100 less.

37. 156 **38.** 433 **39.** 5,212 **40.** 1,082 **41.** 12,617

_____ _____ _____ _____ _____

Write the number that is 1,000 less.

42. 1,092 **43.** 7,214 **44.** 5,131 **45.** 10,673 **46.** 22,194

_____ _____ _____ _____ _____

Find each answer using mental math.

47. $70 - 20 =$ _____   **48.** $400 + 500 =$ _____   **49.** $300 + 600 + 500 =$ _____

50. $800 - 600 =$ _____   **51.** $1,200 - 500 =$ _____   **52.** $4,000 + 9,000 =$ _____

53. $4,200 - 1,200 =$ _____   **54.** $6,300 + 800 =$ _____   **55.** $12,000 + 500 =$ _____

56. Sherri had $1,200 in her savings account. Then she took out $700. The next week she put $900 into her account. The following week, Sherri took out $300. How much is in her account now?

_____

Practice Set 46

Write your answers below or on another piece of paper.

Write the missing number for each Fact Triangle. Then write the family of facts for that triangle.

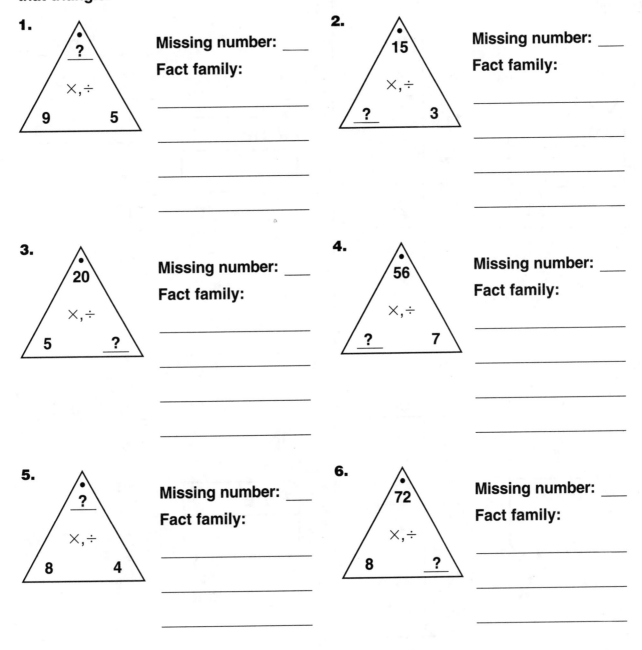

1.

Missing number: ____

Fact family:

_____

_____

_____

_____

2.

Missing number: ____

Fact family:

_____

_____

_____

_____

3.

Missing number: ____

Fact family:

_____

_____

_____

_____

4.

Missing number: ____

Fact family:

_____

_____

_____

_____

5.

Missing number: ____

Fact family:

_____

_____

_____

6.

Missing number: ____

Fact family:

_____

_____

_____

Practice Set 46 *continued*

Use with or after
Lesson 7·2

SRB
203–204

Write your answers below or on another piece of paper.

Write the missing rule and the missing numbers.

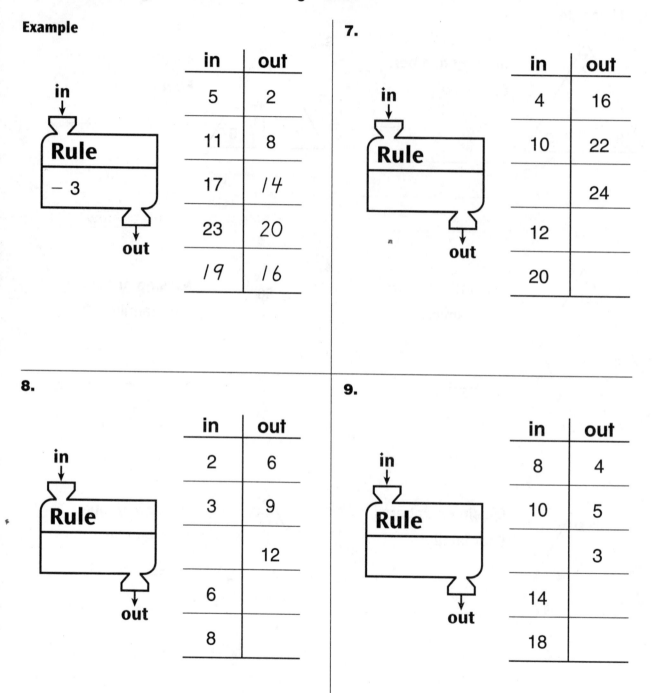

Example

in
↓
Rule
− 3
↓
out

in	out
5	2
11	8
17	14
23	20
19	16

7.

in
↓
Rule
↓
out

in	out
4	16
10	22
	24
12	
20	

8.

in
↓
Rule
↓
out

in	out
2	6
3	9
	12
6	
8	

9.

in
↓
Rule
↓
out

in	out
8	4
10	5
	3
14	
18	

74

Practice Set 47

Write your answers below or on another piece of paper.

Solve.

1. 6 × 6 = _____   **2.** 8 × 3 = _____   **3.** 9 × 6 = _____   **4.** 9 × 9 = _____

5. 7 × 6 = _____   **6.** 8 × 8 = _____   **7.** 8 × 9 = _____   **8.** 8 × 4 = _____

9. 5 × 8 = _____   **10.** 9 × 4 = _____   **11.** 8 × 6 = _____   **12.** 7 × 7 = _____

13. 8 × 6 = _____   **14.** 7 × 8 = _____   **15.** 9 × 5 = _____   **16.** 7 × 9 = _____

Denise invented a game using this gameboard. Answer each question below.

17. How many rows are on Denise's gameboard? _____

How many squares are in each row? _____

18. Write a number model to show the total number

of squares on Denise's gameboard. _____

19. How many of the squares on the gameboard are black? _____

How many are white? _____

20. If 2 markers can be placed on each square of the gameboard, how many markers
can the gameboard hold?

_____

Practice Set 48

Use with or after
Lesson 7·4

SRB
16–17
50–51

Write your answers below or on another piece of paper.

Write a number model. Then solve.

Example	Jerry picked **50** apples. He ate **2** of them. Then he divided the rest of the apples equally into **8** baskets. How many apples did he put in each basket?
	Number model: (50 − 2) ÷ 8 = 6

1. Karen made 3 clay pots Monday and 4 clay pots Tuesday. By the end of the week, she had made 17 pots. How many pots did she make between Wednesday and Friday?

   _____

2. Keneisha had 18 stickers. She put 2 of them on her notebook. She put 4 of them on each of her folders. How many folders did Keneisha have?

   _____

3. Franklin has 82 seashells. He wants to have 100. His friend Mario gave him 7. How many more shells does Franklin need?

   _____

4. Tim needs 24 cupcakes for his birthday party. His mother made 12. Tim has 4 friends who said they will bring the rest. How many cupcakes should each friend bring if all 4 friends bring equal amounts?

   _____

Find each answer.

5. For a picnic, Sharon brought 3 cookies for each of 4 people. How many cookies did she bring in all?

   _____

6. Tom bought 2 packages of postcards. Each package contained 5 postcards. How many postcards did he buy in all?

   _____

76

Practice Set 48 *continued*

Write your answers below or on another piece of paper.

Write each number.

> **Example**
> one million, four hundred ten thousand, five hundred three   **1,410,503**

7. three million, nine hundred fifty-four thousand, six hundred twenty-nine _____

8. nine million, six hundred twenty-one thousand, six hundred eight _____

9. two million, thirty-nine thousand, four hundred ninety-eight _____

10. nine hundred forty-one thousand, eight hundred five _____

11. seven million, three thousand, two hundred eighty _____

12. six million, two hundred nine thousand, four hundred fifty-five _____

13. nine million, eight hundred two _____

14. six million, nine thousand, ten _____

Write the multiplication and division fact family for each group of numbers.

> **Example**   4, 28, 7 $4 \times 7 = 28$
> $7 \times 4 = 28$
> $28 \div 4 = 7$
> $28 \div 7 = 4$

15. 45, 9, 5   **16.** 32, 4, 8   **17.** 20, 4, 80

_____ _____ _____

_____ _____ _____

_____ _____ _____

_____ _____ _____

18. 6, 40, 240   **19.** 10, 6, 60   **20.** 30, 70, 210

_____ _____ _____

_____ _____ _____

_____ _____ _____

_____ _____ _____

Practice Set 49

Use with or after
Lesson 7·6

Write your answers below or on another piece of paper.

Answer each question.

1. How much are 7 [80s]? _____

2. How much are 4 [600s]? _____

3. How much are 2 [2,000s]? _____

4. How much are 3 [400s]? _____

5. Which number multiplied by 3 equals 90? _____

6. Which number multiplied by 4 equals 1,600? _____

7. Which number multiplied by 7 equals 490? _____

Solve each problem.

8. Sharon bought 3 packages of hair bows. There are 5 bows in each package. How many bows did Sharon buy?

_____

9. Joe got 4 packages of stickers as a gift. Each package holds 6 stickers. How many stickers did Joe get?

_____

10. A sheet of stamps has 6 rows. Each row has 3 stamps. How many stamps are on a sheet?

_____

11. Each box of crackers holds 300 crackers. You have no boxes of crackers. How many crackers do you have?

_____

12. Each row of buttons has 6 buttons. You have 1 row of buttons. How many buttons do you have?

_____

13. 5 cakes are each cut into 6 pieces. How many pieces of cake are there?

_____

78

Practice Set 49 continued

Use with or after
Lesson 7·6

SRB
99 100
122

Write your answers below or on another piece of paper.

Measure each line segment to the nearest centimeter. Then tell whether the line segments are *parallel* or *intersecting*.

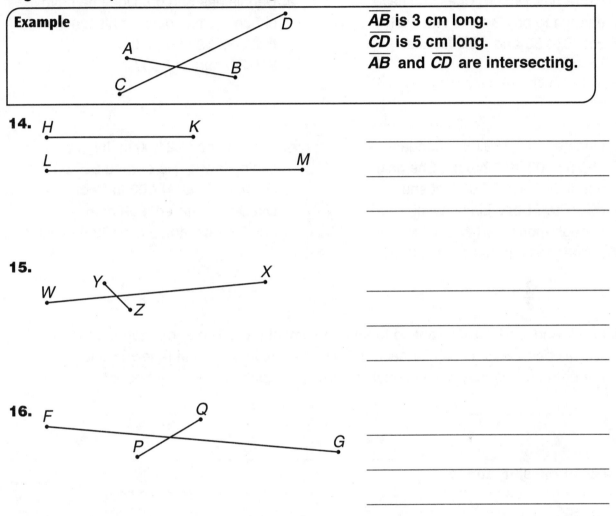

Example

$\overline{AB}$ is 3 cm long.
$\overline{CD}$ is 5 cm long.
$\overline{AB}$ and $\overline{CD}$ are intersecting.

14. H _____ K
L _____ M

_____

_____

_____

15.
W Y X Z

_____

_____

_____

16.
F Q P G

_____

_____

_____

Finish each symmetrical shape according to the line of symmetry.

Example

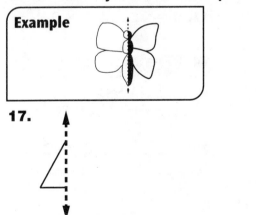

17.

18.

Practice Set 50

Use with or after
Lesson 7·7

SRB
52–53
190–192

Write your answers below or on another piece of paper.

Use estimation to solve each of these problems:

1. Laura has $45.00. Does she have enough to buy 2 skirts that each cost $14.50 and a blouse that costs $17.00?

_____

2. Dennis has $80.00. Does he have enough to buy gifts for his family that cost $23.50, $18.90, $15.95, and $17.50?

_____

3. Linda earned $33.00 in January and $22.00 in February. She paid her sister back $9.00 that she owed her. Does Linda have enough money left to pay for a weekend trip that costs $48.00?

_____

4. Dave earned $24.00 in July by cutting lawns. He earned $29.00 in August and $17.00 in September. Did Dave earn enough money to buy 2 games that cost $33.99 each?

_____

5. Jill, JoAnn, and Jackie want go to an amusement park. The admission fee for each girl will be $22.95. Lunch for each girl will cost $5.50. Each girl received $30.00 for her birthday. Do they have enough money to go to the amusement park?

_____

Find each missing number.

6. $4 = 2 \times$ _____

7. _____ $= 8 \times 0$

8. _____ $\times 5 = 25$

9. $8 \div 1 =$ _____

10. $4 \times$ _____ $= 4$

11. $0 =$ _____ $\times 0$

12. $9 =$ _____ $\div 1$

13. _____ $= 3 \times 6$

14. $27 =$ _____ $\times 9$

15. $4 \times 7 =$ _____

16. $3 \times$ _____ $= 12$

17. _____ $\times 10 = 30$

18. _____ $= 50 \times 1$

19. $1 \div$ _____ $= 1$

20. _____ $\times 6 = 18$

Practice Set 51

Use with or after
Lesson 7·8

SRB
96–101

Write your answers below or on another piece of paper.

Multiply.

1. $7 \times 90 =$ _____   **2.** $40 \times 40 =$ _____   **3.** $10 \times 280 =$ _____   **4.** $5 \times 50 =$ _____

5. $70 \times 6 =$ _____   **6.** $400 \times 8 =$ _____   **7.** $69 \times 10 =$ _____   **8.** $28 \times 10 =$ _____

9. $100 \times 30 =$ _____   **10.** $70 \times 30 =$ _____   **11.** $500 \times 2 =$ _____   **12.** $8 \times 300 =$ _____

Draw the following figures:

13. 2 parallel lines

14. 2 intersecting line segments

15. a right angle

16. 2 rays that form an angle

17. an angle that is smaller than a right angle

18. an angle that shows a half-turn

Find the median for each set of numbers below.

Example	12 16 7 18 13 10
Step 1	Put the numbers in order from least to greatest: 7 10 12 13 16 18
Step 2	Now find the middle value, or the value that has an equal number of values less than and greater than itself.
	The median, or middle value, is between 12 and 13.

19. 27 50 42 18 42

_____

_____

20. 36 9 17 24 28 15 14

_____

_____

21. 82 75 79 81

_____

_____

22. 15 30 19 28 34 11

_____

_____

23. 62 28 55 49 38

_____

_____

24. 83 34 68 58 68 97

_____

_____

Practice Set 51 continued

Use with or after
Lesson 7·8

SRB
33
50–51

Write your answers below or on another piece of paper.

Show each amount of money using the fewest coins and bills possible.

Use $1 s, Q s, D s, N s, and P s.

Example $1.86

$1   Q   Q   Q   D   P

25. 79¢

26. $0.93

27. $1.52

_____ _____ _____

28. 49¢

29. $0.65

30. $2.19

_____ _____ _____

31. $0.98

32. $3.84

33. 59¢

_____ _____ _____

Solve each problem. Then make a ballpark estimate to check that your answer makes sense.

34. Dave caught a fish that was 26 inches long. Kim caught a fish that was 42 inches long. How much longer was Kim's fish?

_____

35. The lowest temperature in Denver one year was 8°F. The highest temperature during that same year was 96°F. What was the difference between the two temperatures?

_____

36. In the morning, the temperature was 24°F. By 2:00 in the afternoon, the temperature was 47°F. How much had the temperature risen?

_____

37. Larry planted a bush that was 38 centimeters tall. Three months later the bush was 51 centimeters tall. How much had the bush grown?

_____

Practice Set 52

Write your answers below or on another piece of paper.

Use the following information to answer the questions below:

A school cafeteria can spend $1.50 on each student per lunch. One hamburger costs the school $1.50. One hot dog, however, costs the school only $0.50.

1. How many hot dogs can replace 1 hamburger? _____

2. How many hot dogs can replace 2 hamburgers? _____

3. How many hot dogs can replace 50 hamburgers? _____

4. How many hot dogs can replace 400 hamburgers? _____

Write the number that has ...

5. 1 hundred-thousand
4 tens
5 ten-thousands
7 ones
9 thousands
3 hundreds

_____

6. 7 ten-thousands
2 ones
9 hundreds
0 thousands
3 hundred-thousands
9 tens

_____

7. 4 hundreds
8 ones
5 hundred-thousands
2 tens
9 thousands
0 ten-thousands

_____

8. 6 thousands
7 ten-thousands
6 ones
8 hundreds
4 tens
5 hundred-thousands

_____

Practice Set 52 *continued*

Use with or after
Lesson 7·9

SRB
150–151
154–156

Write your answers below or on another piece of paper.

Match each description with the correct square or rectangle below. Write the letter that identifies the square or rectangle.

9. a rectangle with a perimeter of 18 units _____

10. a square with an area of 25 square units _____

11. a rectangle with an area of 10 square units _____

12. a rectangle with a perimeter of 10 units _____

13. a square with a perimeter of 20 units _____

14. a square that has the same number
for its perimeter and its area _____

15. a rectangle that has an area of 14 square units _____

16. a rectangle that has a perimeter of 14 units _____

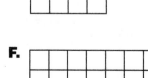

A.

B.

C.

D.

E.

F.

Practice Set 53

**Use with or after
Lesson 8·1**

SRB
22–23

Write your answers below or on another piece of paper.

Write the fraction for the shaded part of each picture.

Example 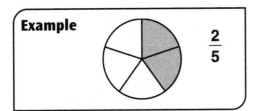 $\dfrac{2}{5}$

1. 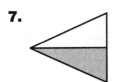 _____

2. 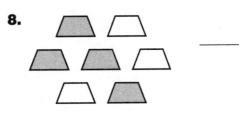 _____

3. 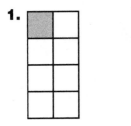 _____

4. 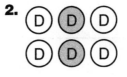 _____

5. _____

6  _____

7. 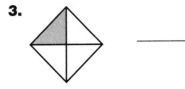 _____

8.  _____

Practice Set 54

Use with or after
Lesson 8·3

SRB
10–12
27–30

Write your answers below or on another piece of paper.

Find the missing numbers on each number line.

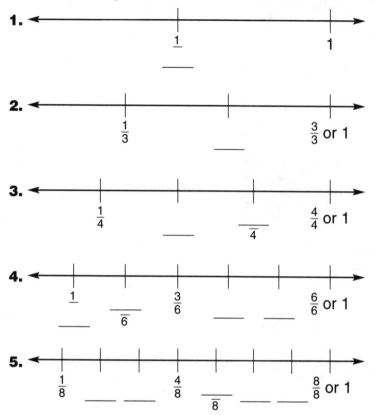

1. $\dfrac{1}{\quad}$ 1

2. $\dfrac{1}{3}$ $\dfrac{\quad}{\quad}$ $\dfrac{3}{3}$ or 1

3. $\dfrac{1}{4}$ $\dfrac{\quad}{\quad}$ $\dfrac{\quad}{4}$ $\dfrac{4}{4}$ or 1

4. $\dfrac{1}{\quad}$ $\dfrac{\quad}{6}$ $\dfrac{3}{6}$ $\dfrac{\quad}{\quad}$ $\dfrac{\quad}{\quad}$ $\dfrac{6}{6}$ or 1

5. $\dfrac{1}{8}$ $\dfrac{\quad}{\quad}$ $\dfrac{\quad}{\quad}$ $\dfrac{4}{8}$ $\dfrac{\quad}{8}$ $\dfrac{\quad}{\quad}$ $\dfrac{\quad}{\quad}$ $\dfrac{8}{8}$ or 1

Find each missing number. You can use the number lines above to help you.

Example $1 = \dfrac{6}{6}$

6. $\dfrac{1}{2} = \dfrac{\quad}{8}$

7. $\dfrac{4}{4} = \dfrac{\quad}{3}$

8. $\dfrac{\quad}{6} = \dfrac{2}{3}$

9. $\dfrac{\quad}{8} = 1$

10. $\dfrac{6}{8} = \dfrac{\quad}{4}$

11. $\dfrac{1}{\quad} = \dfrac{2}{6}$

12. $\dfrac{3}{3} = \dfrac{8}{\quad}$

13. $\dfrac{\quad}{8} = \dfrac{1}{4}$

14. $\dfrac{\quad}{2} = \dfrac{3}{\quad} = \dfrac{4}{8} = \dfrac{\quad}{4}$

Practice Set 55

Use with or after
Lesson 8·4

SRB
22–24

Write your answers below or on another piece of paper.

Write as many numbers as you can for the fractional parts shown in each picture.

Example

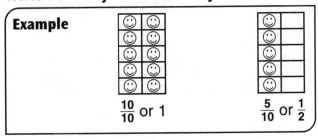

$\frac{10}{10}$ or 1 $\frac{5}{10}$ or $\frac{1}{2}$

$\frac{8}{8}$ or 1

1.

_____

2.

_____

$\frac{6}{6}$ or 1

3.

_____

4.

_____

$\frac{5}{5}$ or 1

5.

_____

6.

_____

$\frac{4}{4}$ or 1

7.

_____

8.

_____

87

Practice Set 55 *continued*

Use with or after
Lesson 8·4

SRB
143–145

Measure each object to the nearest half-inch.

Example

inches

about $2\frac{1}{2}$ inches

9. _____

inches

10. _____

inches

11. _____

inches

Find each product.

12. $1 \times 8 =$ _____

13. $0 \times 5 =$ _____

14. $15 \times 1 =$ _____

15. $24 \times 0 =$ _____

16. $1 \times 37 =$ _____

17. $145 \times 0 =$ _____

Practice Set 56

Write your answers below or on another piece of paper.

Write >, < or =.

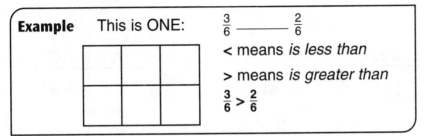

Example   This is ONE: $\frac{3}{6}$ ——— $\frac{2}{6}$

< means *is less than*

> means *is greater than*

$\frac{3}{6}$ > $\frac{2}{6}$

1. This is ONE:

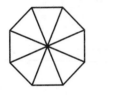

$\frac{1}{2}$ ——— $\frac{2}{4}$ $\frac{3}{4}$ ——— $\frac{1}{4}$

2. This is ONE:

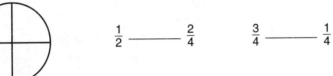

$\frac{5}{8}$ ——— $\frac{3}{4}$ $\frac{1}{2}$ ——— $\frac{4}{8}$

3. This is ONE:

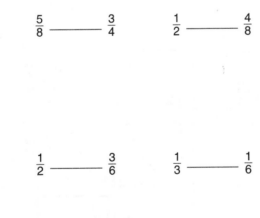

$\frac{1}{2}$ ——— $\frac{3}{6}$ $\frac{1}{3}$ ——— $\frac{1}{6}$

4. This is ONE:

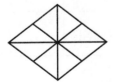

$\frac{3}{8}$ ——— $\frac{1}{4}$ $\frac{2}{8}$ ——— $\frac{1}{4}$

89

Practice Set 56 *continued*

Write your answers below or on another piece of paper.

Write the missing numbers in the tables.

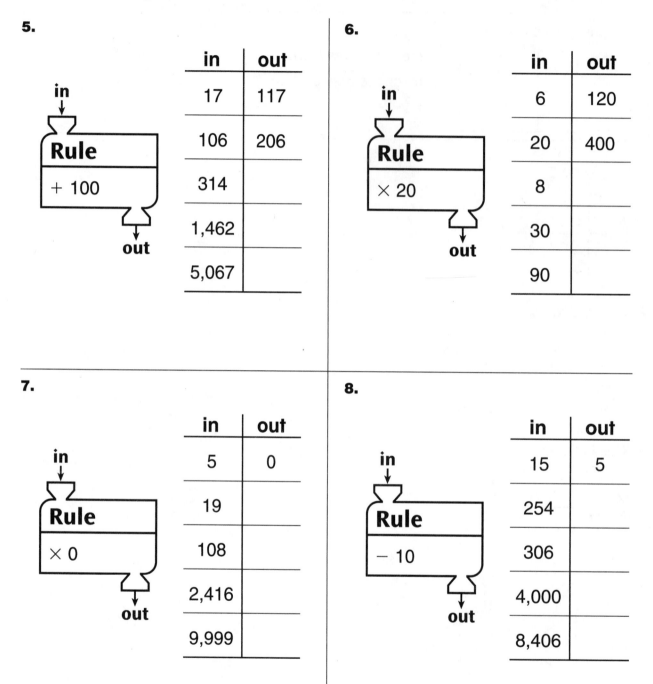

5.

in
↓
Rule
+ 100
out

in	out
17	117
106	206
314	
1,462	
5,067	

6.

in
↓
Rule
× 20
out

in	out
6	120
20	400
8	
30	
90	

7.

in
↓
Rule
× 0
out

in	out
5	0
19	
108	
2,416	
9,999	

8.

in
↓
Rule
− 10
out

in	out
15	5
254	
306	
4,000	
8,406	

Practice Set 57

Write your answers below or on another piece of paper.

Write both a fraction and a mixed number to match each picture.

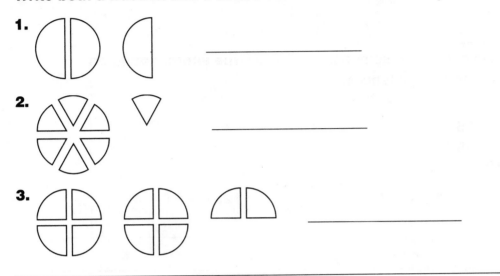

1. _____

2. _____

3. _____

Find the missing number for each Fact Triangle. Then write the fact family for that triangle.

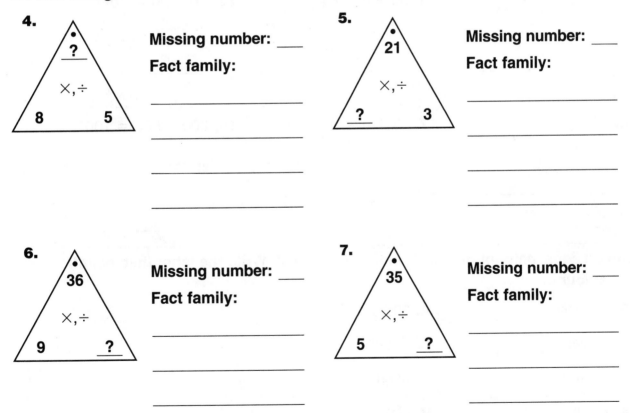

4.

? ×,÷ 8   5

Missing number: ___

Fact family:

_____

_____

_____

_____

5.

21 ×,÷ ?   3

Missing number: ___

Fact family:

_____

_____

_____

_____

6.

36 ×,÷ 9   ?

Missing number: ___

Fact family:

_____

_____

_____

_____

7.

35 ×,÷ 5   ?

Missing number: ___

Fact family:

_____

_____

_____

_____

Practice Set 57 *continued*

Use with or after
Lesson 8·6

SRB
35
50–51

Write your answers below or on another piece of paper.

Complete the number models.

8. $(4 \times 3) + 12 =$ _____

9. $40 = (24 + 36) -$ _____

Use the numbers in each addition sentence below to write another addition sentence and two subtraction sentences.

Example	$19 + 6 = 25$
	$6 + 19 = 25$
	$25 - 6 = 19$
	$25 - 19 = 6$

10. $8 + 6 = 14$

_____

_____

_____

11. $5 + 9 = 14$

_____

_____

_____

12. $7 + 34 = 41$

_____

_____

_____

13. $10 + 7 = 17$

_____

_____

_____

14. $56 + 8 = 64$

_____

_____

_____

15. $40 + 30 = 70$

_____

_____

_____

16. $100 + 4 = 104$

_____

_____

_____

17. $93 + 6 = 99$

_____

_____

_____

18. $200 + 500 = 700$

_____

_____

_____

Find an equal amount of money in the second list. Write the letter that identifies that amount.

19. $\frac{1}{100}$ dollar _____

A. $0.05

20. $\frac{1}{5}$ quarter _____

B. $\frac{1}{10}$ dollar

21. quarter _____

C. penny

22. $\frac{1}{2}$ dollar _____

D. 25¢

23. 10¢ _____

E. 50¢

24. $0.75 _____

F. $\frac{3}{4}$ dollar

Practice Set 58

Use with or after
Lesson 8·7

SRB
14–15
27

Write your answers below or on another piece of paper.

Solve each problem.

1. Sharon brought 12 apples to the picnic. After the picnic, 2 apples were left. What fraction of the apples were eaten?

_____

2. Dave spent 5 days at camp. What fraction of a week did Dave spend at camp?

_____

3. Dorothy bought 10 yards of ribbon. She used 2 yards to wrap packages. What fraction of the ribbon did she use?

_____

4. Glenda had $15. She spent $9 on a book. What fraction of her money did Glenda spend on the book?

_____

5. For a party, a huge sandwich was cut into 25 pieces. After the party, 5 pieces were left. What fraction of the sandwich was eaten? What fraction of the sandwich was not eaten?

_____

6. A vase of flowers has 6 red roses, 6 yellow roses, and 12 white roses. What fraction of the roses are yellow? What fraction of the flowers are white?

_____

Make your own name-collection box for each number. Include +, −, ×, and ÷ at least once in each box. Include at least 8 different names for each number.

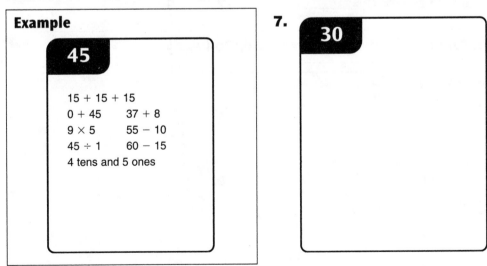

Example

45

15 + 15 + 15
0 + 45   37 + 8
9 × 5   55 − 10
45 ÷ 1   60 − 15
4 tens and 5 ones

7.

30

Practice Set 59

Use with or after
Lesson 9·1

SRB
22–24
52–53

Write your answers below or on another piece of paper.

Solve each problem.

1. a. 200 [400s] _____

   b. 200 × 400 = _____

2. a. 300 [500s] _____

   b. 300 × 500 = _____

3. How many 500s are in 2,000? _____

4. How many 300s are in 3,000? _____

5. How many 200s are in 2,000? _____

Solve each problem. Circle the square products.

6. 7 × 4 = _____   **7.** 4 × 4 = _____   **8.** 6 × 6 = _____

9. 9 × 9 = _____   **10.** 8 × 8 = _____   **11.** 7 × 8 = _____

12. 6 × 9 = _____   **13.** 7 × 7 = _____   **14.** 9 × 8 = _____

15. 5 × 5 = _____   **16.** 8 × 6 = _____   **17.** 7 × 9 = _____

Write a mixed number for each fraction. Draw pictures to help you.

Example $\frac{6}{4}$ ⊕ ◖ $1\frac{2}{4}$

18. $\frac{3}{2}$ _____   **19.** $\frac{6}{5}$ _____   **20.** $\frac{7}{3}$ _____

Practice Set 60

Write your answers below or on another piece of paper.

Solve the following problems mentally:

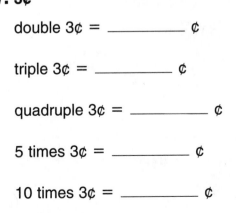

Carton   **Tray**

1. How many eggs are in 2 cartons? _____

2. How many eggs are in 1 tray? _____

3. How many eggs are in half a tray? _____

4. How many eggs are in 3 cartons? _____

5. How many eggs are in half a carton? _____

6. Which is more, 4 cartons or 1 tray? _____

Find each missing number.

Example	**2 gloves**
	double 2 gloves = **4 gloves**
	triple 2 gloves = **6 gloves**
	quadruple 2 gloves = **8 gloves**
	5 times 2 gloves = **10 gloves**
	10 times 2 gloves = **20 gloves**

7. 3¢

double 3¢ = _____ ¢

triple 3¢ = _____ ¢

quadruple 3¢ = _____ ¢

5 times 3¢ = _____ ¢

10 times 3¢ = _____ ¢

8. 4 inches

double 4 in. = _____ in.

triple 4 in. = _____ in.

quadruple 4 in. = _____ in.

5 times 4 in. = _____ in.

10 times 4 in. = _____ in.

95

Practice Set 60 *continued*

Write your answers below or on another piece of paper.

Write the fraction for the shaded part of the picture in two different ways.

Example 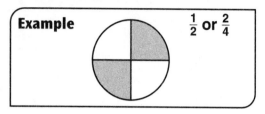 $\frac{1}{2}$ or $\frac{2}{4}$

9.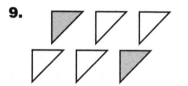

_____

10.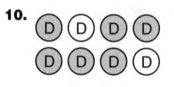

_____

11.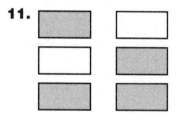

_____

12.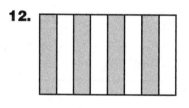

_____

13.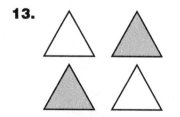

_____

14.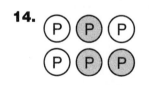

_____

Practice Set 61

Use with or after
Lesson 9·4

Write your answers below or on another piece of paper.

Multiply. Then *check* your answers using a calculator.

Example	29
	× 3
3 [20s]	60
3 [9s]	27
60 + 27	**87**

1. 51
  × 6

2. 84
  × 5

3. 17
  × 8

4. 206
  × 9

5. 419
  × 4

For each problem below, write a number model. Then find the missing numbers.

6. Donna puts 6 pears in each bag. She has 32 pears.
How many bags does she fill?

_____ ÷ _____ → _____ R _____

Donna fills _____ bags.

_____ pears are left over.

7. 21 signs are shared equally by 4 classrooms.
How many signs does each classroom get?

_____ ÷ _____ → _____ R _____

Each classroom gets _____ signs.

_____ signs are left over.

Practice Set 62

Use with or after
Lesson 9·5

SRB
22–24
52–53

Write your answers below or on another piece of paper.

Solve each number story using the grocery store sign.

```
┌──────────────────────────────┐
│      ⌐ Special ⌐             │
│  Peaches    $0.50  each       │
│  Apples     $0.39  each       │
│  Pears      $0.61  each       │
│        (No tax)               │
└──────────────────────────────┘
```

1. Sandra has $2.00. Can she buy 5 apples? _____

   How much money will she have left? _____

2. How much money does Kenny need to buy 4 pears? _____

3. How much does it cost to buy 6 pears and 3 apples? _____

Multiply. Circle each square product.

4. $7 \times 9 =$ _____

5. $8 \times 8 =$ _____

6. $6 \times 7 =$ _____

7. $4 \times 8 =$ _____

8. $5 \times 5 =$ _____

9. $9 \times 9 =$ _____

10. $6 \times 6 =$ _____

11. $7 \times 7 =$ _____

12. $8 \times 9 =$ _____

13. $8 \times 7 =$ _____

14. $6 \times 8 =$ _____

15. $6 \times 6 =$ _____

Change each mixed number to a fraction. Draw pictures to help you.

Example	$1\frac{2}{3}$		$\frac{5}{3}$

16. $2\frac{1}{2}$ _____

17. $3\frac{3}{4}$ _____

18. $1\frac{3}{6}$ _____

Practice Set 62 *continued*

Use with or after
Lesson 9·5

SRB
254–260

Write your answers below or on another piece of paper.

Solve each problem.

19. Each red fox weighs 19 pounds. How much do 7 red foxes weigh?

_____

20. Ben spent 25 minutes walking to school. What fraction of an hour is this? (*Hint:* 1 hour = 60 minutes.)

_____

21. Betty has 25 stickers. She wants to share them equally among 3 friends. How many stickers will each friend get? How many stickers will be left over?

_____

22. Ellen had $30.00. She spent $14.00 shopping. What fraction of her money did she spend? What fraction of her money did she NOT spend?

_____

23. Sam's birthday cake was cut into 16 pieces. After his party, 3 pieces were left. What fraction of his cake was left? What fraction of his cake was eaten?

_____

24. Ruth wants to buy 4 computer games that each cost $29.50. About how much money does Ruth need in order to buy all 4 computer games?

_____

25. Jan wants to get her hair cut and buy shampoo. Jan has $25.00. Does she have enough money for a haircut that costs $16.00 and 2 bottles of shampoo that cost $4.25 each?

_____

26. A quilt has 5 yellow squares, 10 blue squares, and 10 green squares. What fraction of the squares are blue? What fraction of the squares are yellow? What fraction of the squares are red?

_____

99

Practice Set 63

Use with or after
Lesson 9·6

SRB
35
37–38

Write your answers below or on another piece of paper.

Read the information below. Then answer each question.

Ted has 36 flowers. He wants to put the flowers into vases. He wants each vase to have the same number of flowers—without any flowers being left over.

1. Can he put the flowers in 1 vase? 2 vases? 3 vases?

If so, how many flowers go in each vase? _____

2. Can he put the flowers in 4 vases? 5 vases? 6 vases?

If so, how many flowers go in each vase? _____

3. Can he put the flowers in 7 vases? 8 vases? 9 vases?

If so, how many flowers go in each vase? _____

4. Can he put the flowers in 10 vases? 11 vases? 12 vases?

If so, how many flowers go in each vase? _____

The factors of 36 are the numbers that can be multiplied by whole numbers to get 36, or the numbers that 36 can be divided by without having remainders.

5. Name the factors of 36. _____

For each number below, give the value of each digit.

Example 48.613	**The 4 means 4 tens.**
	The 8 means 8 ones.
	The 6 means 6 tenths.
	The 1 means 1 hundredth.
	The 3 means 3 thousandths.

6. 295.6 _____

7. 30.48 _____

8. 10.925 _____

_____   _____   _____

_____   _____   _____

_____   _____   _____

Practice Set 63 *continued*

Write your answers below or on another piece of paper.

Write the number family for each Fact Triangle.

Example

△ 210
×, ÷
7   30

7 × 30 = 210
30 × 7 = 210
210 ÷ 7 = 30
210 ÷ 30 = 7

9.

△ 400
×, ÷
50   8

_____

_____

10.

△ 630
×, ÷
90   7

_____

_____

11.

△ 240
×, ÷
3   80

_____

_____

12.

△ 1,000
×, ÷
10   100

_____

_____

13.

△ 3,500
×, ÷
50   70

_____

_____

14.

△ 1,800
×, ÷
20   90

_____

_____

Practice Set 64

Use with or after
Lesson 9·7

SRB
10–12
52–53

Write your answers below or on another piece of paper.

Solve each problem.

1. $56 ÷ 8 = _____

2. $81 ÷ 9 = _____

3. $54 ÷ 6 = _____

4. $150 ÷ 6 = _____

5. $120 ÷ 8 = _____

6. $140 ÷ 7 = _____

7. $122 ÷ 4 = _____

8. $85 ÷ 5 = _____

9. $490 ÷ 7 = _____

Find the missing numbers. Use fractions.

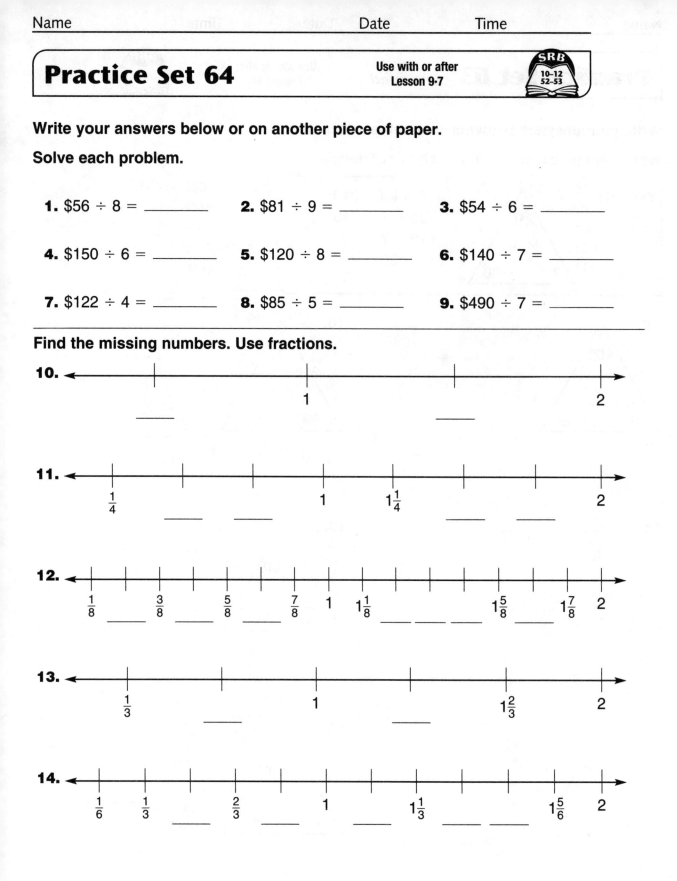

Practice Set 65

Write your answers below or on another piece of paper.

Solve each problem.

1. Ken wants to put 6 ounces of water in each glass. How many glasses can he fill with 42 ounces of water? How many ounces of water will be left over?

2. Lynn wants to cut a 50-inch piece of string into pieces that are each 8 inches long. How many 8-inch pieces can she cut? How many inches of string will be left over?

_____

Write the number that has ...

3. 7 in the tens place
1 in the thousands place
4 in the tenths place
2 in the hundreds place
6 in the ones place

_____

4. 4 in the hundredths place
1 in the tens place
5 in the ones place
7 in the tenths place
9 in the hundreds place

_____

5. 0 in the tenths place
9 in the thousandths place
2 in the ones place
5 in the tens place
8 in the hundredths place

_____

6. 0 in the hundredths place
6 in the ones place
1 in the tens place
9 in the thousandths place
3 in the tenths place

_____

7. 3 in the hundreds place
9 in the ones place
8 in the tenths place
4 in the tens place
6 in the thousands place
7 in the hundredths place

_____

8. 4 in the tenths place
2 in the thousands place
0 in the tens place
6 in the thousandths place
1 in the ones place
0 in the hundredths place
4 in the hundreds place

_____

103

Practice Set 66

Write your answers below or on another piece of paper.

Use lattice multiplication to solve each problem.

1. 8 × 49 = _____

2. 7 × 359 = _____

3. 6 × 314 = _____

4. 9 × 68 = _____

5. 5 × 456 = _____

6. 7 × 834 = _____

Write a multiplication fact to find the area of each square.

Example	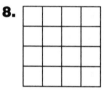	3 × 3 = 9 Area = 9 square units

7.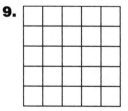

_____

_____

8.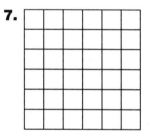

_____

_____

9.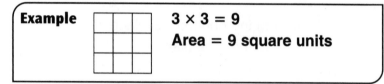

_____

10.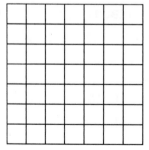

_____

Practice Set 67

Use with or after
Lesson 9·12

SRB
190–192
250–253

Write your answers below or on another piece of paper.

Multiply using the partial-products method. Then use a calculator to check each answer.

Example	78
	× 43
(40 × 70)	2,800
(40 × 8)	320
(3 × 70)	210
(3 × 8)	24
	3,354

1.   29
      × 14

2.   34
      × 51

3.   62
      × 22

4.   44
      × 36

5.   81
      × 53

6.   39
      × 28

7.   76
      × 64

Use estimation to solve each problem.

8. Dina has $25.00. Does she have enough to buy a radio that costs $14.79 and a CD that costs $11.25?

_____

9. Jaime has $40.00. Does he have enough to buy shoes that cost $21.97 and two ties that cost $8.50 each?

_____

10. Janice has $50.00. How many plants can she buy if each plant costs $11.95?

_____

11. Jack has $60.00. How many shirts can he buy if each shirt costs $14.50?

_____

12. Jason has $63.00 in his savings account. He received $35.00 as birthday gifts. Does he have enough money to buy a bike that costs $90.00?

_____

Practice Set 67 *continued*

Write your answers below or on another piece of paper.

Solve each problem.

13. Arthur bought a goldfish for 49¢, a striped fish for $0.72, and fish food for 68¢. How much did Arthur spend?

_____

14. How much change did Arthur receive if he paid for the 3 items with $3.00?

_____

15. Betty wants to buy a dog collar for $3.50, a water dish for $2.79, and a toy bone for $3.49. Can she buy all 3 items with $10.00?

_____

16. How much do the 3 items that Betty wants to buy cost altogether?

_____

Write >, <, or =.

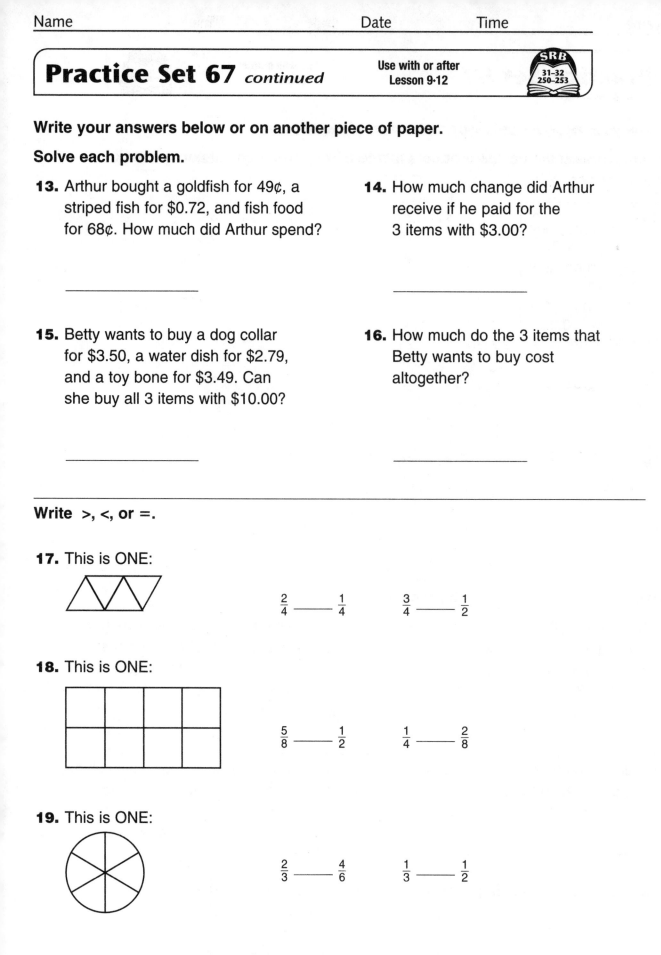

17. This is ONE:

$\frac{2}{4}$ _____ $\frac{1}{4}$ $\frac{3}{4}$ _____ $\frac{1}{2}$

18. This is ONE:

$\frac{5}{8}$ _____ $\frac{1}{2}$ $\frac{1}{4}$ _____ $\frac{2}{8}$

19. This is ONE:

$\frac{2}{3}$ _____ $\frac{4}{6}$ $\frac{1}{3}$ _____ $\frac{1}{2}$

Practice Set 68

Use with or after
Lesson 9·13

SRB
11
171–173

Write your answers below or on another piece of paper.

Write the number and the unit for each problem. Use Celsius temperatures.

Example 4 degrees below zero **−4°C**

1. 25 degrees above zero _____

2. 58 degrees below zero _____

3. zero degrees _____

4. 150 degrees above zero _____

5. 14 degrees below zero _____

6. 100 degrees below zero _____

Which temperature is colder? You can use the thermometer to help you.

Example −6°C or −14°C
  −14°C is below −6°C on the thermometer.
  −14°C is colder than −6°C.

7. 0°C or 5°C

8. 2°C or −20°C

_____   _____

9. 9°C or −9°C

10. −7°C or 0°C

_____   _____

Which temperature is warmer? You can use the thermometer to help you.

Example 24°C or −2°C
  24°C is above −2°C on the thermometer.
  24°C is warmer than −2°C.

11. 0°C or −8°C

12. −98°C or 1°C

_____   _____

13. 15°C or −15°C

14. 12°C or −35°C

_____   _____

°F   °C
210 — — 100
200 —
190 — — 90
180 —
170 — — 80
160 — — 70
150 —
140 — — 60
130 —
120 — — 50
110 —
100 — — 40
90 —
80 — — 30
70 — — 20
60 —
50 — — 10
40 —
30 — — 0
20 —
10 — — −10
0 —
−10 — — −20
−20 — — −30
−30 —
−40 — — −40
−50 —
−60 — — −50
−70 —
−80 — — −60
−90 — — −70
−100 —
−110 — — −80
−120 —
−130 — — −90
−140 —
−150 — — −100

107

Practice Set 68 *continued*

Use with or after
Lesson 9·13

SRB
37–38
250–253

Write your answers below or on another piece of paper.

Complete the list of factors for each number below.

> *The factors of a number* are the numbers that can be multiplied by whole numbers to get that number, or the numbers that a number can be divided by without having remainders.

Example   Factors of 16: 1, ___**2**___, ___**4**___, 8, ___**16**___

15. Factors of 7: _____, 7

16. Factors of 18: _____, 2, _____, _____, _____, 18

17. Factors of 36: 1, _____, _____, 4, _____, 9, _____, _____, 36

18. Factors of 50: _____, 2, _____, _____, 25, _____

Solve each problem.

19. Jerry went swimming 19 days in June. What fraction of the days in June did Jerry go swimming? (*Hint:* June has 30 days.)

20. Sarah spent 20 minutes eating breakfast. What fraction of an hour did she spend eating breakfast? What fraction of an hour did she NOT spend eating breakfast?

_____   _____

21. Sam and two friends shared a pizza cut into 8 pieces. Sam ate 1 piece, and each of his friends ate 2 pieces.

What fraction of the pizza did Sam eat? _____

What fraction of the pizza did each friend eat? _____

What fraction of the pizza was left over? _____

Practice Set 69

Use with or after
Lesson 10·1

SRB
140 146

Write your answers below or on another piece of paper.

Find each missing number. Use fractions.

> 1 meter = 10 decimeters   1 yard = 3 feet
> 1 meter = 100 centimeters   1 yard = 36 inches
> 1 decimeter = 10 centimeters   1 foot = 12 inches
> 1 centimeter = 10 millimeters

1. _____ yard = 12 inches

2. _____ meter = 8 decimeters

3. 1 foot = _____ yard

4. 50 centimeters = _____ meter

5. 9 inches = _____ foot

6. _____ centimeter = 3 millimeters

7. _____ yard = 2 feet

8. 9 centimeters = _____ meter

9. _____ meter = 2 decimeters

10. _____ decimeter = 3 centimeters

11. 5 inches = _____ foot

12. _____ meter = 3 centimeters

Find each answer. You can draw pictures or use counters.

13. There are 12 students taking swimming lessons. $\frac{1}{3}$ of them are third graders. How many are third graders?

14. The pet store has 10 dogs for sale. Half of the dogs are collies. How many of the dogs are collies?

15. Karen drew a picture of 8 flags. She colored $\frac{1}{4}$ of the flags orange.

How many flags did she color orange? _____

What fraction of the flags did she NOT color orange? _____

109

Practice Set 69 *continued*

Write your answers below or on another piece of paper.

For each ruler, find the distance between the two points.

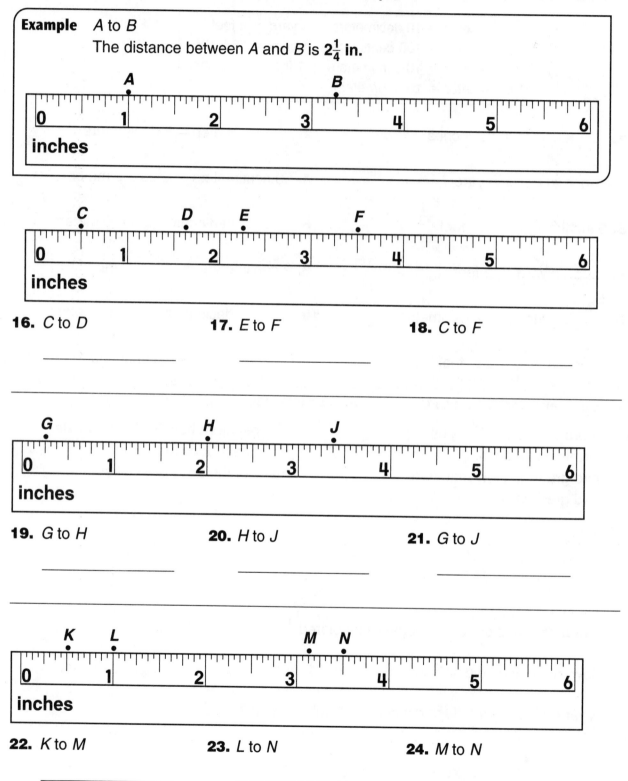

Example   *A* to *B*
The distance between *A* and *B* is $2\frac{1}{4}$ in.

16. *C* to *D*

17. *E* to *F*

18. *C* to *F*

_____

_____

_____

19. *G* to *H*

20. *H* to *J*

21. *G* to *J*

_____

_____

_____

22. *K* to *M*

23. *L* to *N*

24. *M* to *N*

_____

_____

_____

110

Practice Set 70

Write your answers below or on another piece of paper.

Find the volume (V) of each box. Each cube stands for 1 cubic centimeter.

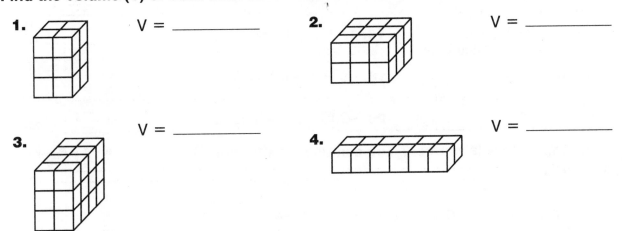

1. V = _____

2. V = _____

3. V = _____

4. V = _____

Tell which unit you would use to measure each item. Choose from inch, foot, yard, and mile.

> **Example**   the length of a pencil
> **Unit: inch**

5. the length of your math book _____

6. the length of a paper clip _____

7. the distance between Chicago and St. Louis _____

8. the length of a football field _____

9. the width of your hand _____

10. the width of a room _____

11. the width of your foot _____

12. the width of a park _____

13. the distance traveled in a car after one hour _____

14. the height of a dog _____

Practice Set 70 *continued*

Write your answers below or on another piece of paper.

Which temperature is colder? You can use the thermometer to help you.

15. 0°F or –18°F _____   **16.** 12°F or –12°F _____

17. 0°F or 6°F _____   **18.** –8°F or –18°F _____

19. 5°F or 15°F _____   **20.** –23°F or –32°F _____

Which temperature is warmer? You can use the thermometer to help you.

21. 6°C or 36°C _____   **22.** –14°C or –45°C _____

23. 0°C or –10° C _____   **24.** 12°C or –12°C _____

25. 16°C or –37°C _____   **26.** 20°C or 0°C _____

Solve each problem. You can use the thermometer to help you.

27. One January morning, the temperature was –18°F. By noon, the temperature had risen to 4°F. How many degrees had the temperature risen?

_____

28. One June morning, the temperature was 18°C. By 2:00 in the afternoon, the temperature had risen to 34°C. How many degrees had the temperature risen?

_____

29. On Tuesday, the high temperature was –10°C. On Friday, the high temperature was 4°C. How many degrees warmer was the high temperature on Friday?

_____

Practice Set 71

**Use with or after
Lesson 10-4**

162–166

Write your answers below or on another piece of paper.

Read the scale and record the weight.

1.

_____

2.

_____

3.

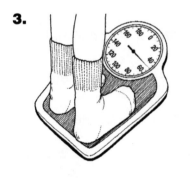

_____

4.

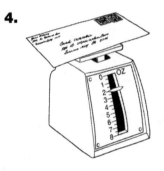

_____

5.

_____

Practice Set 71 *continued*

Use with or after
Lesson 10·4

SRB
37 40

Write your answers below or on another piece of paper.

Find the missing numbers. Use fractions.

1 meter = 10 decimeters	1 yard = 3 feet
1 meter = 100 centimeters	1 yard = 36 inches
1 decimeter = 10 centimeters	1 foot = 12 inches
1 centimeter = 10 millimeters	

Example   3 centimeters = ___$\frac{3}{100}$___ meter

6. 1 inch = _____ foot

7. _____ meter = 1 centimeter

8. 1 foot = _____ yard

9. _____ decimeter = 1 centimeter

10. _____ meter = 1 decimeter

11. 6 centimeters = _____ meter

12. 12 inches = _____ foot

13. _____ centimeter = 1 millimeter

14. 2 feet = _____ yard

15. 70 centimeters = _____ meter

16. _____ yard = 24 inches

17. 9 decimeters = _____ meter

Find the factors for each number listed below.

The factors of a number are the numbers that can be multiplied by whole numbers to get that number, or the numbers that a number can be divided by without having remainders.

18. 9

19. 20

20. 35

_____   _____   _____

21. 60

22. 8

23. 51

_____   _____   _____

24. 32

25. 45

26. 21

_____   _____   _____

114

Practice Set 72

Use with or after
Lesson 10·6

SRB
160–161

Write your answers below or on another piece of paper.

Tell which unit you would use to measure each item. Choose from gallon, quart, pint, cup, ounce, and tablespoon.

1. amount of water you drink with dinner _____

2. container of milk that you buy at the store_____

3. amount of water in a bathtub _____

4. amount of juice in a can from a vending machine _____

5. amount of syrup on pancakes _____

6. container of orange juice that you buy at the store _____

7. amount of water in an eyedropper _____

8. amount of water in a swimming pool _____

9. amount of cream in a cup of coffee _____

10. amount of lemonade needed to serve 4 people _____

Solve each problem. Circle the square products.

Unit
African lions

11. $9 \times 9 =$ _____

12. $8 \times 7 =$ _____

13. $6 \times 8 =$ _____

14. $9 \times 6 =$ _____

15. $7 \times 7 =$ _____

16. $8 \times 9 =$ _____

17. $7 \times 6 =$ _____

18. $8 \times 8 =$ _____

19. $7 \times 9 =$ _____

20. 400 [800s] = _____

21. 100 [700s] = _____

22. 200 [500s] = _____

Practice Set 73

Write your answers below or on another piece of paper.

Find the mean for each data set below.

Example	9 5 7 6 3
Step 1	Find the total of the numbers in the data set. $9 + 5 + 7 + 6 + 3 = 30$
Step 2	Count the numbers in the data set. There are 5 numbers in all.
Step 3	Divide the total by 5. $30 \div 5 = 6$ **The mean is 6.**

1. 7 2 5 6

_____

2. 5 4 2 5 6 2

_____

3. 12 8 7 10 13

_____

4. 9 5 6 10 10 11 12

_____

Find each missing number.

1 mile (mi)	=	1,760 yards (yd)
1 mile (mi)	=	5,280 feet (ft)
1 yard (yd)	=	3 feet (ft)
1 yard (yd)	=	36 inches (in.)
1 foot (ft)	=	12 inches (in.)

5. _____ feet = 2 yards

6. 18 inches = _____ ft _____ in.

7. 3 yd 2 ft = _____ ft

8. 9 ft = _____ yd _____ in.

9. 2 miles = _____ yards

10. 10,560 feet = _____ miles

11. 75 in. = _____ yd _____ in.

12. 3 ft 9 in. = _____ in.

13. 5,500 ft = _____ mi _____ ft

14. 15 ft = _____ yd

Practice Set 73 *continued*

Use with or after
Lesson 10·7

SRB
86–87

Write your answers below or on another piece of paper.

Use the bar graph to answer each question below.

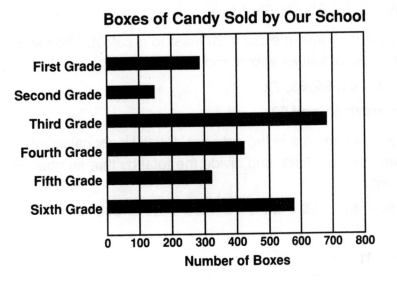

Boxes of Candy Sold by Our School

Number of Boxes

15. Which grade sold the fewest boxes of candy?

_____

16. Which grade sold the most boxes of candy?

_____

17. What is the range, or difference, between the highest and lowest numbers on the graph?

_____

18. Which grade sold about twice as many boxes of candy as the second grade sold?

_____

19. Put the grades in order from the grade that sold the most boxes to the grade that sold the fewest boxes.

_____

20. How many more boxes of candy would the fourth grade have to sell in order to reach 500 boxes?

_____

Name _____ Date _____ Time _____

Practice Set 74

Write your answers below or on another piece of paper.

Find the median and the mean for each data set below.

> **Example**   58, 63, 65, 49, 53, 65, 58, 72, 65, 61
>
> To find the **median,** put the numbers in order from least to greatest. The median is the number with an equal number of values above and below it.
>
> **49, 53, 58, 58, 61, 63, 65, 65, 65, 72**
>
> **The median is between 61 and 63.**
>
> To find the **mean,** add all the numbers in the data set. (*Hint:* Use a calculator.) Then count how many numbers are in the set, and divide the total by that number. Round to the nearest whole number.
>
> **58 + 63 + 65 + 49 + 53 + 65 + 58 + 72 + 65 + 61 = 609**
>
> **609 ÷ 10 = 60.9**
>
> **The mean is about 61.**

1. 195, 204, 198, 187, 200, 193, 205, 187, 182

_____

_____

2. 929, 905, 917, 933, 904, 913, 905, 901, 908, 922

_____

_____

3. 397, 395, 400, 403, 397, 382, 410, 406, 395, 397

_____

_____

4. 1,111; 1,083; 1,102; 1,075; 1,096; 1,114; 1,111; 1,075

_____

_____

5. 15,270; 15,400; 15,230; 15,320; 15,290; 15,405; 15,300; 15,240; 15,320

_____

_____

Practice Set 74 *continued*

Use with or after
Lesson 10-8

SRB
18–20
160–161

Write your answers below or on another piece of paper.

Match each measurement in the first list with an equal measurement in the second list. Write the letter that identifies that equal measurement.

1 gallon	= 4 quarts
1 quart	= 2 pints
1 pint	= 2 cups
1 cup	= 8 fluid ounces

6. 2 gallons _____   **A.** 4 pints

7. 5 cups _____   **B.** $\frac{1}{16}$ gallon

8. 4 fluid ounces _____   **C.** 24 fluid ounces

9. $2\frac{1}{2}$ pints _____   **D.** 6 pints

10. $\frac{1}{2}$ gallon _____   **E.** 5 cups

11. 3 pints _____   **F.** $\frac{1}{2}$ cup

12. 3 quarts _____   **G.** 6 quarts

13. $1\frac{1}{2}$ gallons _____   **H.** 8 quarts

14. 3 cups _____   **I.** 40 fluid ounces

15. 1 cup _____   **J.** 6 cups

Write the value of the underlined digit in each number.

16. 479,2̲14   **17.** 289.4̲6̲   **18.** 7̲8,432

_____   _____   _____

19. 3̲.289   **20.** 1o̲8.27   **21.** 1̲29,568

_____   _____   _____

22. 1̲,045,618   **23.** 157̲,214   **24.** 40.67̲1

_____   _____   _____

25. 3̲57,490   **26.** 2,046.0̲3   **27.** 92,491.04̲5

_____   _____   _____

Practice Set 75

Use with or after
Lesson 10·10

SRB
70–78

Write your answers below or on another piece of paper.

The frequency table below shows the number of boxes of cards sold by the third grade to raise money for the school library. Use a calculator to help you answer each question.

Number of Children	Number of Boxes Sold
1	### ### ### ###
2	### ### //
3	### ///
4	### //
5	### ###
6	### /
7	###
8	////

1. What is the total number of children who sold cards?

   _____

2. How many boxes of cards were sold in all?

   _____

3. Find the **mode,** or the number that occurs most often, of the number of boxes of cards sold.

   _____

4. Find the **median** number of boxes of cards sold.

   _____

5. Find the **mean** number of boxes of cards sold.

   _____

Solve each problem.

6. Jason had 25 quarters. He put 7 of them in his bank. What fraction of the quarters did Jason put in his bank?

   _____

7. Becky spends 5 hours each day at school. What fraction of the day does Becky spend at school? (*Hint:* A day has 24 hours.)

   _____

8. Bryan has 4 history videos, 5 science videos, and 3 adventure videos. What fraction of his videos are history? What fraction of his videos are adventure?

   _____

Practice Set 75 *continued*

Use with or after
Lesson 10·10

SRB
22–23
250–253

Write your answers below or on another piece of paper.

Draw and shade shapes to show each fraction.

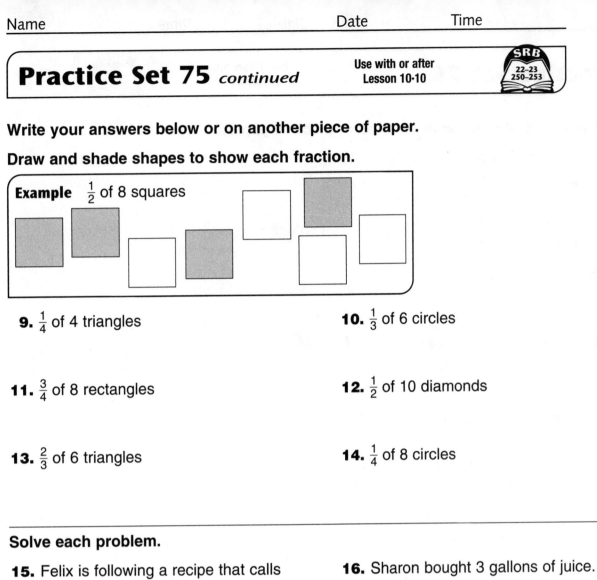

Example $\frac{1}{2}$ of 8 squares

9. $\frac{1}{4}$ of 4 triangles

10. $\frac{1}{3}$ of 6 circles

11. $\frac{3}{4}$ of 8 rectangles

12. $\frac{1}{2}$ of 10 diamonds

13. $\frac{2}{3}$ of 6 triangles

14. $\frac{1}{4}$ of 8 circles

Solve each problem.

15. Felix is following a recipe that calls for 3 cups of milk. How many cups of milk does he need to double the recipe?

16. Sharon bought 3 gallons of juice. There are 4 quarts in 1 gallon. How many quarts of juice did Sharon buy?

17. At the end of the school year, Lisa weighed 62 pounds. She had gained 6 pounds during the school year. How much did Lisa weigh at the beginning of school?

18. Sarah rode her bike 18 kilometers Monday and 25 kilometers Tuesday. How many kilometers did she ride her bike in all those two days?

19. A garden in the shape of a square measures 3 meters on each side. How many meters of fencing would you need to put fencing around the entire garden?

Practice Set 76

Write your answers below or on another piece of paper.

Follow the directions.

1. Plot these points:

A: (1,1)	B: (3,6)
C: (6,3)	D: (7,0)
E: (5,6)	F: (7,9)
G: (10,6)	H: (3,9)

2. Draw the following line segments: $\overline{AB}, \overline{BC}, \overline{CA}$

What shape did you make? _____

3. Draw the following line segments: $\overline{DE}, \overline{EF}, \overline{FG}, \overline{GD}$

What shape did you make? _____

4. Which shape is symmetrical? _____

5. Measure the following line segments to the nearest centimeter: $\overline{AB}, \overline{DG}, \overline{FG}, \overline{CE}$

_____

Find the volume (V) of each box. Each cube stands for 1 cubic inch.

6. V = _____

7. V = _____

8. V = _____

9. V = _____

122

Practice Set 77

Use with or after
Lesson 11·3

Write your answers below or on another piece of paper.

Imagine that each of the following containers is tipped over onto a table.

1.

500 Pennies

How many HEADS? _____ How many TAILS? _____

2.

200 counters

How many *black* sides faceup? _____

How many *white* sides faceup? _____

3.

50 buttons

How many *front* sides faceup? _____

How many *back* sides faceup? _____

Solve each division problem. If the problem has a remainder, write that amount after the letter *R*.

Example $100 \div 9 \rightarrow$ **11 R1**

4. $81 \div 9 \rightarrow$ _____

5. $54 \div 6 \rightarrow$ _____

6. $72 \div 8 \rightarrow$ _____

7. $56 \div 8 \rightarrow$ _____

8. $48 \div 6 \rightarrow$ _____

9. $36 \div 6 \rightarrow$ _____

10. $35 \div 4 \rightarrow$ _____

11. $200 \div 6 \rightarrow$ _____

12. $95 \div 4 \rightarrow$ _____

13. $17 \div 6 \rightarrow$ _____

14. $3,000 \div 5 \rightarrow$ _____

15. $332 \div 10 \rightarrow$ _____

Practice Set 78

Use with or after
Lesson 11·5

SRB
83–85
92–94

Write your answers below or on another piece of paper.

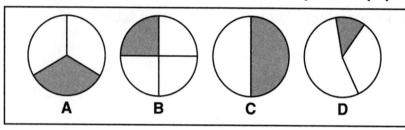

A   B   C   D

On which of the spinners above ...

1. are you equally likely to land on the shaded part or the white part?

_____

2. are you likely to land on the shaded part about $\frac{1}{4}$ of the time?

_____

3. are you twice as likely to land on a white part?

_____

4. are you likely to land on a white part about $\frac{3}{4}$ of the time?

_____

5. are you likely to land on the shaded part about $\frac{1}{3}$ of the time?

_____

Find the mean of each data set below.

To *find the mean,* add all the numbers in the data set. Then count how many numbers are in the set, and divide the total by that number. Round to the nearest whole number.

6. 349, 756, 821, 444, 348, 259 _____

7. 3,500; 3,511; 3,487; 3,548 _____

8. 28, 34, 56, 54, 76, 89, 21, 13, 49, 112 _____

9. 1,001; 1,012; 998; 799; 804; 1,030 _____

Practice Set 78 *continued*

Write your answers below or on another piece of paper.

Write the missing numbers.

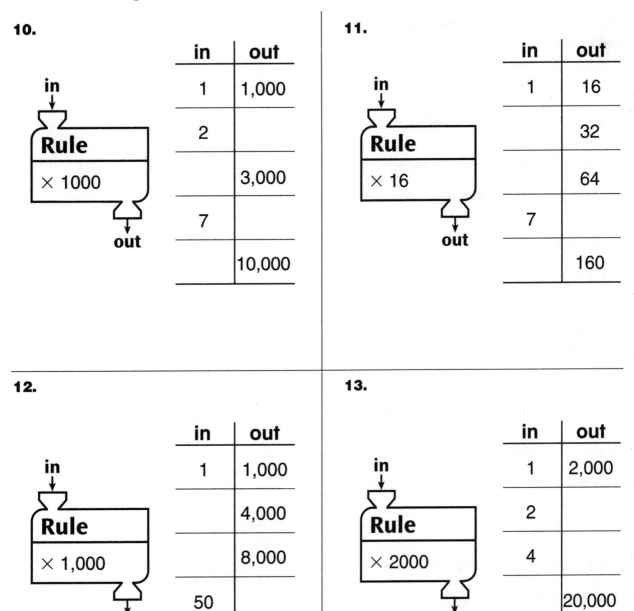

10.

in	out
1	1,000
2	
	3,000
7	
	10,000

Rule × 1000

11.

in	out
1	16
	32
	64
7	
	160

Rule × 16

12.

in	out
1	1,000
	4,000
	8,000
50	
100	

Rule × 1,000

13.

in	out
1	2,000
2	
4	
	20,000
7	

Rule × 2000

Practice Set 79

Write your answers below or on another piece of paper.

Make the following predictions:

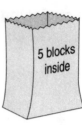

5 blocks inside

1. You make 10 random draws. *You draw …*
- blue 4 times
- red 4 times
- yellow 2 times

Predict the colors of the 5 blocks in the bag. _____

Tell what fraction of the blocks is NOT yellow. _____

2. You make 25 random draws. *You draw …*
- orange 16 times
- purple 9 times

Predict the colors of the 5 blocks in the bag. _____

3. You make 50 random draws. *You draw …*
- red 19 times
- blue 21 times
- white 10 times

Predict the colors of the 5 blocks in the bag. _____

Tell what fraction of the blocks are white. _____

4. You make 50 random draws. *You draw …*
- blue 26 times
- red 24 times

Predict the colors of the 6 blocks in the bag. _____

5. You make 40 random draws. *You draw …*
- pink 6 times
- orange 13 times
- red 7 times
- green 14 times

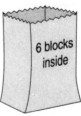

6 blocks inside

Predict the colors of the 6 blocks in the bag. _____

Tell what fraction of the blocks are pink. _____

Practice Set 79 *continued*

Use with or after
Lesson 11·6

SRB
54 55

Write your answers below or on another piece of paper.

Find the missing number for each Fact Triangle. Then write the number family for that triangle.

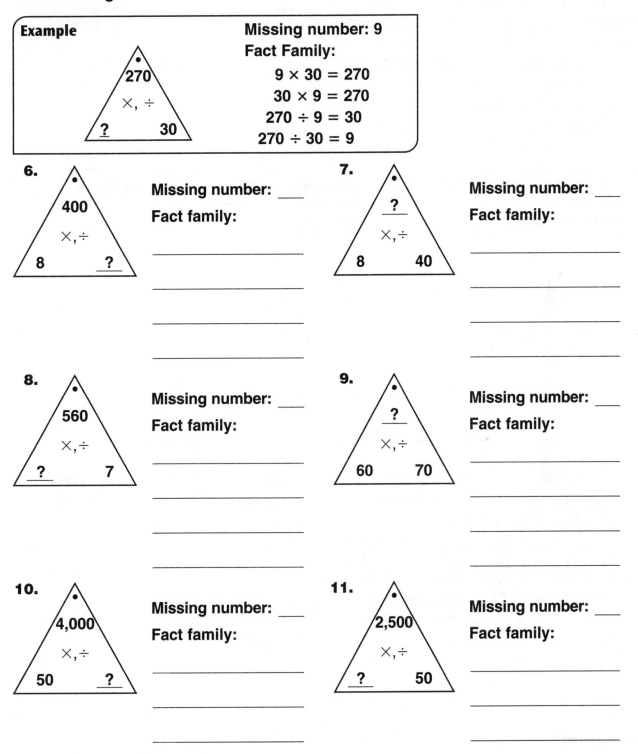

Example

270

×, ÷

?   30

Missing number: 9
Fact Family:
     9 × 30 = 270
     30 × 9 = 270
     270 ÷ 9 = 30
     270 ÷ 30 = 9

6.

400

×, ÷

8   ?

Missing number: ___

Fact family:

_____

_____

_____

_____

7.

?

×, ÷

8   40

Missing number: ___

Fact family:

_____

_____

_____

_____

8.

560

×, ÷

?   7

Missing number: ___

Fact family:

_____

_____

_____

_____

9.

?

×, ÷

60   70

Missing number: ___

Fact family:

_____

_____

_____

_____

10.

4,000

×, ÷

50   ?

Missing number: ___

Fact family:

_____

_____

_____

_____

11.

2,500

×, ÷

?   50

Missing number: ___

Fact family:

_____

_____

_____

_____

Practice Set 80

Use with or after
Lesson 11·7

SRB
22–24

Write your answers below or on another piece of paper.

Add numbers to the chart and find the total for each column. Then answer the questions.

There are 5 third grade classes at Lincoln Elementary.
- Room 101 has 12 boys and 12 girls.
- Room 102 has 14 boys and 12 girls.
- Room 103 has 13 boys and 13 girls.
- Room 104 has 11 boys and 13 girls.
- Room 105 has 12 boys and 11 girls.

1.

Number of Third Graders at Lincoln Elementary School		
Room Number	**Boys**	**Girls**
101		
102		
103		
104		
105		
TOTALS		

2. Use fractions to tell about how many of the third grade students are girls and about how many are boys.

_____

3. There are about 100 *second* graders at the school. Predict how many are boys and how many are girls.

_____

4. There are about 130 *fourth* graders at the school. Predict how many are girls and how many are boys.

_____

5. Can you predict whether a new student in the third grade will be a boy or a girl?

_____

128